Debra A. Pellegrino Smith, Ed.D.

INDIVIDUALISM
OLD AND NEW

INDIVIDUALISM
OLD AND NEW

JOHN DEWEY

GREAT BOOKS IN PHILOSOPHY

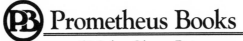 Prometheus Books
59 John Glenn Drive
Amherst, New York 14228-2197

Published 1999 by Prometheus Books
59 John Glenn Drive, Amherst, New York 14228–2197,
716–691–0133, ext. 207. FAX: 716–564–2711.
WWW.PROMETHEUSBOOKS.COM

Library of Congress Cataloging-in-Publication Data

Dewey, John, 1859–1952.
    Individualism old and new / John Dewey.
        p.   cm. — (Great books in philosophy)

    ISBN 1–57392–693–0 (pbk. : alk. paper)
    1. Individualism. 2. United States—Social conditions.
3. United States—Civilization. I. Title. II. Series.
HM136.D4   1999
306'.0973—dc21   99–10420
                                                              CIP

Printed in the United States of America on acid-free paper.

Additional Titles on Social and Political Philosophy in Prometheus's Great Books in Philosophy Series

Aristotle: *The Politics*

Francis Bacon: *Essays*

Mikhail Bakunin: *The Basic Bakunin: Writings, 1869–1871*

Edmund Burke: *Reflections on the Revolution in France*

John Dewey: *Freedom and Culture*

G. W. F. Hegel: *The Philosophy of History*

G. W. F. Hegel: *Philosophy of Right*

Thomas Hobbes: *The Leviathan*

Sidney Hook: *Paradoxes of Freedom*

Sidney Hook: *Reason, Social Myths, and Democracy*

John Locke: *Second Treatise on Civil Government*

Niccolo Machiavelli: *The Prince*

Karl Marx: *The Poverty of Philosophy*

Karl Marx and Friedrich Engels: *The Economic and Philosophical Manuscripts of 1844* and *The Communist Manifesto*

Karl Marx (with Friedrich Engels): *The German Ideology,* including *Theses on Feuerbach* and *Introduction to the Critique of Political Economy*

John Stuart Mill: *Considerations on Representative Government*

John Stuart Mill: *On Liberty*

John Stuart Mill: *On Socialism*

John Stuart Mill: *The Subjection of Women*

Friedrich Nietzsche: *Thus Spake Zarathustra*

Thomas Paine: *Common Sense*

Thomas Paine: *Rights of Man*

Plato: *Lysis, Phaedrus,* and *Symposium*

Plato: *The Republic*

Jean-Jacques Rousseau: *The Social Contract*

Mary Wollstonecraft: *A Vindication of the Rights of Men*

Mary Wollstonecraft: *A Vindication of the Rights of Women*

See the back of this volume for a complete list of titles in Prometheus's Great Books in Philosophy and Great Minds series.

JOHN DEWEY was born near Burlington, Vermont, on October 20, 1859. Twenty years later, he graduated from the University of Vermont, after which he taught public school in Pennsylvania and Vermont. Having become interested in philosophical questions while still an undergraduate, Dewey continued his philosophical training at Johns Hopkins University. In 1884 he was awarded a doctorate in philosophy from that institution and soon thereafter he accepted a position in philosophy at the University of Michigan. Except for a one-year appointment as professor of philosophy at the University of Minnesota, Dewey remained at Michigan—serving a five-year term as chairman—until 1894 when he moved with his wife, Alice Chipman, to the University of Chicago and began his tenure as chairman of the philosophy department. It was at Chicago that Dewey received national recognition for his pioneering work in the field of education with the development of his laboratory school in which experimental approaches to teaching were explored. After a falling out with the University of Chicago over the administration of the school, Dewey left in 1904 and accepted a professorship in philosophy at Columbia University.

For the next twenty-six years, Dewey's academic position at Columbia served as a springboard for his many and varied interests—e.g., social questions, politics, education, and public affairs—his national and international reputation found him working with such groups as the American Philosophical Association, the American Association of University Professors (founder and first president), the Teacher's Union, and the American Civil Liberties Union, among others.

Unlike those who consider retirement a time to relax and enjoy the restful pleasures of later life, John Dewey dedicated his remaining years to sorting out the tough social questions facing America and the world. He joined organizations whose goal was to increase public education in the areas of domestic and international politics. One of Dewey's most famous public forums was his participation in the

vii

commission that met in Mexico City to inquire into the charges lev-
eled against Leon Trotsky at his Moscow trial. The commission sub-
sequently found Trotsky innocent of the charges. He was also one of
several colleagues who publicly defended fellow philosopher Bertrand
Russell when Russell was denied a teaching position at the City Col-
lege of New York because of public criticism of his views on marriage
and religion.

In developing his own unique philosophical stance, John Dewey
overcame Hegelian idealism to embrace the pragmatic views of
William James. Dewey's devotion to free inquiry and the scientific
method found him spearheading the intellectual opposition against
the belief that absolute knowledge can be attained in a world of var-
iegated circumstances, discoveries, trailblazing research, and
advances of all kinds. For Dewey, knowledge is not absolute,
immutable, and eternal, but rather relative to the developmental
interaction of man with his world as problems arise to present themselves for solution. This scientific approach, which allows one to
declare the truth of a claim until—and only until—there is negative
evidence sufficient to disconfirm the hypothesis, opens the mind to
the need for a democratic approach to problem solving. Without
cooperation and a rational tolerance for diverse points of view within
a pluralistic community, society has no hope of mature development.

During his ninety-three years, John Dewey authored more than
two dozen books and scores of articles in both scholarly and popular
publications. He is truly America's foremost philosopher, whose work
will influence intellectuals throughout the world for many years to
come.

John Dewey died in New York City on June 1, 1952.

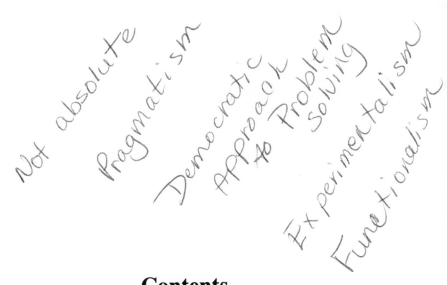

Not absolute

Pragmatism

Democratic

Approach to Problem solving)

Experimentalism

Functionalism

Contents

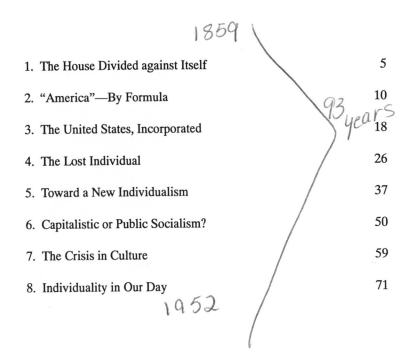

1859

93 years

1952

Individualism, Old and New

Prefatory Note

I am obliged to the courtesy of the editors of the *New Republic* for permission to use material that originally appeared in the columns of that journal and which is now incorporated in connection with considerable new matter, in this volume. It is a pleasure to acknowledge my particular indebtedness to Mr. Daniel Mebane, the treasurer of the *New Republic*, for valuable suggestions and assistance.

Rejected Idealism
→ did not opt
for materialism
yet synthesis

School and Society

Charles Darwin
Origin of Species

① Evolutionary Theory

Transactional

② Young Hegelian
Movement

Thesis -
antithesis -
Synthesis

③ Scientific
Method — Neo (rebirth)
- Enlightment
Perspective

Synthesis the
new Dualism

Child and the
curriculum

Endorsement
of
Democracy

1914
Democracy
in
Education

Old and New
Individualism
to Larger Social
Order

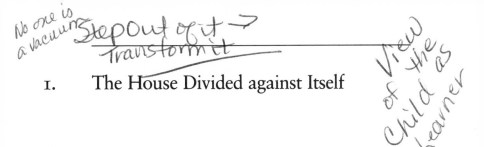

No one is a vacuum *Step out of it →* *Transform it*

View of the as child as learner

1. The House Divided against Itself

It is becoming a commonplace to say that in thought and feeling, or at least in the language in which they are expressed, we are living in some bygone century, anywhere from the thirteenth to the eighteenth, although physically and externally we belong to the twentieth century. In such a contradictory condition, it is not surprising that a report of American life, such as is contained, for example, in *Middletown*, should frequently refer to a "bewildered" or "confused" state of mind as characteristic of us.

Anthropologically speaking, we are living in a money culture. *money culture* Its cult and rites dominate. "The money medium of exchange and the cluster of activities associated with its acquisition drastically condition the other activities of the people." This, of course, is as it should be; people have to make a living, do they not? And for what should they work if not for money, and how should they get goods and enjoyments if not by buying them with money—thus enabling someone else to make more money, and in the end to start shops and factories to give employment to still others, so that they can make more money to enable other people to make more money by selling goods—and so on indefi- *individualism* nitely. So far, all is for the best in the best of all possible cultures: our rugged—or is it ragged?—individualism.

And if the culture pattern works out so that society is divided into two classes, the working group and the business (including professional) group, with two and a half times as many in the former as in the latter, and with the chief ambition of parents in the former class that their children should climb into the latter, that is doubtless because American life offers such unparalleled

[First published as "The House Divided against Itself," in *New Republic* 58 (24 April 1929): 270–71.]

Young children to shape the environment → Transactional Philosophy

opportunities for each individual to prosper according to his virtues. If few workers know what they are making or the meaning of what they do, and still fewer know what becomes of the work of their hands—in the largest industry of Middletown perhaps one-tenth of one percent of the product is consumed locally—this is doubtless because we have so perfected our system of distribution that the whole country is one. And if the mass of workers live in constant fear of loss of their jobs, this is doubtless because our spirit of progress, manifest in change of fashions, invention of new machines and power of overproduction, keeps everything on the move. Our reward of industry and thrift is so accurately adjusted to individual ability that it is natural and proper that the workers should look forward with dread to the age of fifty or fifty-five, when they will be laid on the shelf.

All this we take for granted; it is treated as an inevitable part of our social system. To dwell on the dark side of it is to blaspheme against our religion of prosperity. But it is a system that calls for a hard and strenuous philosophy. If one looks at what we do and what happens, and then expects to find a theory of life that harmonizes with the actual situation, he will be shocked by the contradiction he comes upon. For the situation calls for assertion of complete economic determinism. We live as if economic forces determined the growth and decay of institutions and settled the fate of individuals. Liberty becomes a well-nigh obsolete term; we start, go, and stop at the signal of a vast industrial machine. Again, the actual system would seem to imply a pretty definitely materialistic scheme of value. Worth is measured by ability to hold one's own or to get ahead in a competitive pecuniary race. "Within the privacy of shabby or ambitious houses, marriage, birth, child-rearing, death, and the personal immensities of family life go forward. However, it is not so much these functional urgencies of life that determine how favorable this physical necessity shall be, but the extraneous detail of how much money the father earns." The philosophy appropriate to such a situation is that of struggle for existence and survival of the economically fit. One would expect the current theory of life, if it reflects the actual situation, to be the most drastic Darwinism. And, finally, one would anticipate that the personal traits most prized would be clear-sighted vision of personal advantage and resolute ambition to secure it at any human cost. Sentiment and sympathy would be at the lowest discount.

It is unnecessary to say that the current view of life in Middle-town, in Anytown, is nothing of this sort. Nothing gives us Americans the horrors more than to hear that some misguided creature in some low part of the earth preaches what we prac-tise—and practise much more efficiently than anyone else—namely, economic determinism. Our whole theory is that man plans and uses machines for his own humane and moral pur-poses, instead of being borne wherever the machine carries him. Instead of materialism, our idealism is probably the loudest and most frequently professed philosophy the world has ever heard. We praise even our most successful men, not for their ruthless and self-centered energy in getting ahead, but because of their love of flowers, children, and dogs, or their kindness to aged relatives. Anyone who frankly urges a selfish creed of life is everywhere frowned upon. Along with the disappearance of the home, and the multiplication of divorce in one generation by 600 percent, there is the most abundant and most sentimental glorification of the sacredness of home and the beauties of con-stant love that history can record. We are surcharged with altru-ism and bursting with desire to "serve" others.

These are only a few of the obvious contradictions between our institutions and practice on one hand, and our creeds and theories on the other, contradictions which a survey of any of our Middletowns reveals. It is not surprising that the inhabitants of these towns are bewildered, uneasy, restless, always seeking something new and different, only to find, as a rule, the same old thing in a new dress. It may all be summed up, perhaps, by say-ing that nowhere in the world at any time has religion been so thoroughly respectable as with us, and so nearly totally discon-nected from life. I hesitate to dwell on the revelation that this book gives of "religious" life in Middletown. The glorification of religion as setting the final seal of approval on pecuniary success, and supplying the active motive to more energetic struggle for such success, and the adoption by the churches of the latest devices of the movies and the advertiser, approach too close to the obscene. Schooling is developed to the point where more pupils reach the high school than in other lands; and one-half of the pupils in the last years of the high school think that the first chapters of the Hebrew Scriptures give a more accurate account of the origin and early history of man than does science, and only one-fifth actively dissent. If the investigation had been made when a certain questionnaire was distributed among our school children, it is likely that the usual percentage of youth would have recorded their belief that Harding was the greatest man in the world. In another way, the whole story is told in brief when one contrasts what is actually happening to family life and the complete secularization of daily activities with a statement from the pulpit that "the three notable words in the English language are mother, home and heaven," a remark that would certainly pass unquestioned in any representative American audience.

It makes little difference whether one selects important or trivial aspects of the contradiction between our life as we outwardly live it and our thoughts and feelings—or what we at least say are our beliefs and sentiments. The significant question is: What is the cause of this split and contradiction? There are those, of course, who attribute it to the fact that people being, generally speaking, morons and boobs, they must be expected to act out the parts to which they are assigned. The "explanation" does not take us very far, even if one accepts it. The particular forms that the alleged boobery takes are left quite unaccounted for. And the more one knows of history, the more one comes to believe that traditions and institutions count more than native capacity or incapacity in explaining things. It is evident enough that the rapid industrialization of our civilization took us unawares. Being mentally and morally unprepared, our older creeds have become ingrowing; the more we depart from them in fact, the more loudly we proclaim them. In effect we treat them as magic formulae. By repeating them often enough we hope to ward off the evils of the new situation, or at least to prevent ourselves from seeing them—and this latter function is ably performed by our nominal beliefs.

With an enormous command of instrumentalities, with possession of a secure technology, we glorify the past, and legalize and idealize the *status quo*, instead of seriously asking how we are to employ the means at our disposal so as to form an equitable and stable society. This is our great abdication. It explains how and why we are a house divided against itself. Our tradition, our heritage, is itself double. It contains in itself the ideal of equality of opportunity and of freedom for all, without regard to birth and status, as a condition for the effective realization of that equality. This ideal and endeavor in its behalf once constituted our essential Americanism; that which was prized as the note of a new world. It is the genuinely spiritual element of our tradition. No one can truthfully say that it has entirely disappeared. But its promise of a new moral and religious outlook has not been attained. It has not become the well-spring of a new intellectual consensus; it is not (even unconsciously) the vital source of any distinctive and shared philosophy. It directs our politics only spasmodically, and while it has generously provided schools it does not control their aims or their methods.

Meanwhile our institutions embody another and older tradition. Industry and business conducted for money profit are nothing new; they are not the product of our own age and culture; they come to us from a long past. But the invention of the machine has given them a power and scope they never had in the past from which they derive. Our law and politics and the incidents of human association depend upon a novel combination of the machine and money, and the result is the pecuniary culture characteristic of our civilization. The spiritual factor of our tradition, equal opportunity and free association and intercommunication, is obscured and crowded out. Instead of the development of individualities which it prophetically set forth, there is a perversion of the whole ideal of individualism to conform to the practices of a pecuniary culture. It has become the source and justification of inequalities and oppressions. Hence our compromises, and the conflicts in which aims and standards are confused beyond recognition.

2. "America"—By Formula

We have heard a good deal of late years of class-consciousness. The phrase "nation-conscious" does not happen to be current, but present-day nationalism is an exacerbated expression of it in fact. A still more recent manifestation might be called "culture-consciousness" or "civilization-consciousness." Like class-consciousness and nationalism, it assumes an invidious form; it is an exponent—and a coefficient—of conflict between groups. The war and its consequences may not have produced in our own country a consciousness of "Americanism" as a distinctive mode of civilization but they have definitely had that effect among the intellectual elite of Europe.

Americanism as a form of culture did not exist, before the war, for Europeans. Now it does exist and as a menace. In reaction and as a protest, there is developing, at least among literary folk, the consciousness of a culture which is distinctively European, something which is precious and whose very existence is threatened by an invasion of a new form of barbarism issuing from the United States. Acute hostility to a powerful alien influence is taking the place of complacent ignoring of what was felt to be negligible. It would take a wider knowledge than mine to list even the titles of books and articles coming yearly from the presses of Europe whose burden is the threat of "America" to the traditional culture of Europe.

I am not concerned here with the European side of the matter. Most social unifications come about in response to external pressure. The same is likely to be true of a United States of Europe. If the ideal is approximated in reality, it will probably be as a protective reaction to the economic and financial hegemony of the United States of America. The result would probably be a good

[First published as "'America'—By Formula," in *New Republic* 60 (18 September 1929): 117–19.]

thing for Europe, and thus unwittingly we should serve one good purpose, internationally speaking. But it is, in the end, no great consolation to know that in losing our own soul we have been a means of helping save the soul of some one else. Just what is the America whose picture is forming in the minds of European critics?

Some of the writers are ignorant as well as bitter. These may be neglected. Others are intelligent, as well-informed as any foreigner can be about a foreign country, and not devoid of sympathy. Moreover their judgments agree not only with one another but with the protests of native-born dissenters. For convenience and because of the straightforward intelligence of its author, I take as a point of departure, the description of the American type of mind and character presented by Mueller Freienfels.[1] His treatment is the fairer because he understands by "American" a type of mind that is developing, from like causes, all over the world, and which would have emerged in time in Europe, even if there were no geographical America, although its development in the rest of the world has been accelerated and intensified by the influence of this country.

As far as any actual American is true to the type that is proclaimed to be *the* American, he should be thrilled by the picture that is drawn of him. For we are told that the type is a genuine mutation in the history of culture; that it is new, the product of the last century, and that it is stamped with success. It is transforming the external conditions of life and thereby reacting on the psychical content of life; it is assimilating other types of itself and re-coining them. No world-conquest, whether that of Rome or Christendom, compares with that of "Americanism" in extent or effectiveness. If success and quantity are in fact the standards of the "American," here are admissions that will content his soul. From the standpoint of the type depicted, he is approved; and what do adverse criticisms matter?

But either the type is not yet so definitely fixed as is represented, or else there are individual Americans who deviate from type. For there are many who will have reserves in their admira-

1. *Mysteries of the Soul*, translated from the German by Bernard Miall; New York, Knopf, 1929. It may be well to add, in view of the title, that there is nothing occult nor obscurantist about the book. By "soul" is meant "the manifold living reciprocal reactions between the self and the universe."

tion of the picture that is presented. Of course the dissenters may be, as the European critics say, impotent sports, fish out of water and affected with nostalgia for the European tradition. Nevertheless, it is worth while to raise the question as to whether the American type, supposing there is to be one, has as yet taken on definitive form. First, however, what are alleged to be the characteristics of the type?

Fundamentally, they spring from impersonality. The roots of the intellect are unconscious and vital, in instincts and emotions. In America, we are told, this subconsciousness is disregarded; it is suppressed or is subordinated to conscious rationality, which means that it is adapted to the needs and conditions of the external world. We have "intellect," but distinctly in the Bergsonian sense; mind attuned to the conditions of action upon matter, upon the world. Our emotional life is quick, excitable, undiscriminating, lacking in individuality and in direction by intellectual life. Hence the "externality and superficiality of the American soul"; it has no ultimate inner unity and uniqueness—no true personality.

The marks and signs of this "impersonalization" of the human soul are quantification of life, with its attendant disregard of quality; its mechanization and the almost universal habit of esteeming technique as an end, not as a means, so that organic and intellectual life is also "rationalized"; and, finally, standardization. Differences and distinctions are ignored and overridden; agreement, similarity, is the ideal. There is not only absence of social discrimination but of intellectual; critical thinking is conspicuous by its absence. Our pronounced trait is mass suggestibility. The adaptability and flexibility that we display in our practical intelligence when dealing with external conditions have found their way into our souls. Homogeneity of thought and emotion has become an ideal.

Quantification, mechanization and standardization: these are then the marks of the Americanization that is conquering the world. They have their good side; external conditions and the standard of living are undoubtedly improved. But their effects are not limited to these matters; they have invaded mind and character, and subdued the soul to their own dye. The criticism is familiar; it is so much the burden of our own critics that one is never quite sure how much of the picture of foreign critics is

Homogenous Society

drawn from direct observation and how much from native novels and essays that are not complacent with the American scene. This fact does not detract from the force of the indictment; it rather adds to it, and raises the more insistently the question of what our life means.

I shall not deny the existence of these characteristics, nor of the manifold evils of superficialism and externalism that result in the production of intellectual and moral mediocrity. In the main these traits exist and they characterize American life and are already beginning to dominate that of other countries. But their import is another thing than their existence. Mueller Freienfels is intelligent enough to acknowledge that they are transitional rather than final. He recognizes that the forces are so intrinsic that it is foolish to rebel against them and lament the past. "The question is how we are to pass through them and transcend them." It is this note which distinguishes his appraisal from so many others.

In reply to the question, one may at least say that we are still in an early stage of the transition. Anything that is at most but a hundred years old has hardly had time to disclose its meaning in the slow secular processes of human history. And it may be questioned whether even our author has not sometimes succumbed to the weakness of lesser critics in treating the passing symptoms as inherent characters. I do not have in mind here an "optimistic" appeal to future time and its possibilities. I rather wish to raise the question as to how many of the defects and evils that are supposed to belong to the present order are in fact projections into it of a departing past order.

Strength, power, is always relative, not absolute. Conquest is an exhibition of weakness in the conquered as well as of strength in the conqueror. Transitions are out of something as well as into something; they reveal a past as well as project a future. There must have been something profoundly awry in the quality, spirituality and individualized variety of the past, or they would not have succumbed as readily as we are told they are doing to the quantification, mechanization and standardization of the present. And the defective and perverse elements have certainly not been displaced. They survive in the present. Present conditions give these factors an opportunity to disclose themselves. They are not now kept under and out of sight. Their overt manifestation is not a cheering spectacle. But as long as they did not show themselves on a scale large enough to attract notice, they could not be dealt with. I wonder very much whether many of the things that are objected to in the present scene—and justly so— are not in fact revelations of what the older type of culture covered up, and whether their perceptible presence is not to be credited rather than debited to the forces that are now active.

It is possible of course to argue—as Keyserling for example seems to do—that the new or American order signifies simply that the animal instincts of man have been released, while the older European tradition kept them in disciplined subjection to something higher, called with pleasing vagueness "spirituality." The suspicion that suppression is not solution is not confined to America. Undue and indiscriminate greediness in the presence of accessible food may be a symptom of previous starvation rather than an inevitable exhibition of the old Adam. A culture whose tradition rests on depreciation of the flesh and on making a sharp difference between body and mind, instinct and reason, practice and theory, may have wrought corruption of flesh and degeneration of spirit. It would take a degree of wisdom no one possesses to tell just what, in the undesirable features of the present, is a reflection of an old but not as yet transformed system of life and thought, and how much is a genuine product of the new forces.

One thing seems to be reasonably certain. The prized and vaunted "individuality" of European culture that is threatened by the leveling standardization and uniformity of the American type was a very limited affair. If one were to retort in kind, one could ask how much share in it was had by the peasant and proletarian. And it is much more than a retort to say that a peasantry and proletariat which has been released from intellectual bondage will for a time have its revenge. Because there is no magic in democracy to confer immediately the power of critical discrimination upon the masses who have been outside any intellectual movement, and who have taken their morals and their religion from an external authority above them—an authority which science is destroying—it does not follow that the ineptitude of the many is the creation of democracy.

Take one instance—the present interest in technique, and the domination of the "American type" by technique. It will hardly be argued I suppose that the mere absence of technique—intelligent means and methods for securing results—is itself a mark of an intrinsically desirable civilization. Nor is it surprising that the discovery of the actuality and potentiality of technique in all branches of human life should have an immediately intoxicating effect. What is called the American mentality is characterized by this discovery, and by the exaggerations that come with the abruptness of the discovery. There is much to be said against quantification and standardization. But the discovery of competent technique stands on a different level. The world has not suffered from absence of ideals and spiritual aims anywhere nearly as much as it has suffered from absence of means for realizing the ends which it has prized in a literary and sentimental way. Technique is still a novelty in most matters, and like most novelties is played with for a while on its own account. But it will be used for ends beyond itself sometime; and I think that interest in technique is precisely the thing which is most promising in our civilization, the thing which in the end will break down devotion to external standardization and the mass-quantity ideal. For its application has not gone far as yet; and interest in it is still largely vicarious, being that, so to say, of the spectator rather than of naturalization in use. In the end, technique can only signify emancipation of individuality, and emancipation on a broader scale than has obtained in the past.

In his most hopeful anticipation of a future to which we may be moving, Freienfels calls attention to the fact that the impoverishment of the individual is accompanied, even now, by an enrichment of community resources. Collectively, present society, he says, is marked by a power over nature and by intellectual resource and power exceeding that of the classic Athenian and the man of the Renaissance. Why is it that this collective enrichment does not operate to elevate correspondingly the life of individuals? This question he does not ask. Failure to consider it constitutes to my mind the chief failure of critics whether foreign or native. Our materialism, our devotion to money making and to having a good time are not things by themselves. They are the product of the fact that we live in a money culture; of the fact that our technique and technology are controlled by interest in private profit. There lies the serious and fundamental defect of our civilization, the source of the secondary and induced evils to which so much attention is given. Critics are dealing with symptoms and effects. The evasion of fundamental economic causes by critics both foreign and native seems to me to be an indication of the prevalence of the old European tradition, with its disregard for the body, material things and practical concerns. The development of the American type, in the sense of the critics, is an expression of the fact that we have retained this tradition and the economic system of private gain on which it is based, while at the same time we have made an independent development of industry and technology that is nothing short of revolutionary. When our critics deal with this issue instead of avoiding it there will be something really doing.

Until the issue is met, the confusion of a civilization divided against itself will persist. The mass development, which our European critics tell us has submerged individuality, *is* the product of a machine age; in some form it will follow in all countries from the extension of a machine technology. Its immediate effect has been, without doubt, a subjection of certain types of individuality. As far as individuality is associated with aristocracy of the historic type, the extension of the machine age will presumably be hostile to individuality in its traditional sense all over the world. But the strictures of our European critics only define the issue touched upon in the previous chapter. The problem of constructing a new individuality consonant with the objective conditions under which we live is the deepest problem of our times.

There are two "solutions" that fail to solve. One of these is the method of avoidance. This course is taken as far as it is assumed that the only valid type of individuality is that which holds over from the ages that anteceded machine technology and the democratic society it creates. The course that is complementary to the method of escape springs from assumption that the present situation is final; that it presents something inherently ultimate and fixed. Only as it is treated as transitive and moving, as material to be dealt with in shaping a later outcome, only, that is, as it is treated as a *problem*, is the idea of any solution genuine and relevant. We may well take the formula advanced by European critics as a means of developing our consciousness of some of the conditions of the problem. So regarded, the problem is seen to be essentially that of creation of a new individualism as significant for modern conditions as the old individualism at its best was for

world problem

its day and place. The first step in further definition of this problem is realization of the collective age which we have already entered. When that is apprehended, the issue will define itself as utilization of the realities of a corporate civilization to validate and embody the distinctive moral element in the American version of individualism: Equality and freedom expressed not merely externally and politically but through personal participation in the development of a shared culture.

3. The United States, Incorporated

It was not long ago that it was fashionable for both American and foreign observers of our national scene to sum up the phenomena of our social life under the title of "individualism." Some treated this alleged individualism as our distinctive achievement; some critics held that it was the source of our backwardness, the mark of a relatively uncivilized estate. To-day both interpretations seem equally inept and outmoded. Individualism is still carried on our banners and attempts are made to use it as a war-cry, especially when it is desired to defeat governmental regulation of any form of industry previously exempt from legal control. Even in high quarters, rugged individualism is praised as the glory of American life. But such words have little relation to the moving facts of that life.

There is no word which adequately expresses what is taking place. "Socialism" has too specific political and economic associations to be appropriate. "Collectivism" is more neutral, but it, too, is a party-word rather than a descriptive term. Perhaps the constantly increasing role of corporations in our economic life gives a clue to a fitting name. The word may be used in a wider sense than is conveyed by its technical legal meaning. We may then say that the United States has steadily moved from an earlier pioneer individualism to a condition of dominant corporateness. The influence business corporations exercise in determining present industrial and economic activities is both a cause and a symbol of the tendency to combination in all phases of life. Associations tightly or loosely organized more and more define the opportunities, the choices and the actions of individuals.

I have said that the growth of legal corporations in manufac-

[First published as "Individualism, Old and New. I: The United States, Incorporated," in *New Republic* 61 (22 January 1930): 239–41.]

turing, transportation, distribution and finance is symbolic of the development of corporateness in all phases of life. The era of trust-busting is an almost forgotten age. Not only are big mergers the order of the day, but popular sentiment now looks upon them with pride rather than with fear. Size is our current measure of greatness in this as in other matters. It is not necessary to ask whether the opportunity for speculative manipulation for the sake of private gain, or increased public service at a lower cost, is the dominant motive. Personal motives hardly count as productive causes in comparison with impersonal forces. Mass production and mass distribution inevitably follow in the wake of an epoch of steam and electricity. These have created a common market, the parts of which are held together by intercommunication and interdependence; distance is eliminated and the tempo of action enormously accelerated. Aggregated capital and concentrated control are the contemporary responses.

Political control is needed, but the movement cannot be arrested by legislation. Witness the condition of nearly innocuous desuetude of the Sherman Anti-Trust Act. Newspapers, manufacturing plants, utilities supplying light, power and local transportation, banks, retail stores, theaters and the movies, have all joined in the movement toward integration. General Motors, the American Telegraph and Telephone Company, United States Steel, the rapid growth of chain-store systems, combinations of radio companies with companies controlling theaters all over the country, are familiar facts. Railway consolidations have been slowed up by politics and internal difficulties, but few persons doubt that they, too, are coming. The political control of the future to be effective must take a positive instead of negative form.

For the forces at work in this movement are too vast and complex to cease operation at the behest of legislation. Aside from direct evasions of laws, there are many legal methods of carrying the movement forward. Interlocking directorates, interpurchase of stocks by individuals and corporations, grouping into holding companies, investing companies with enough holdings to sway policies, effect the same end as do direct mergers. It was stated at a recent convention of bankers that eighty per cent of the capitalization of all the banks of the country is now in the hands of twelve financial concerns. It is evident that virtual control of the other twenty per cent, except for negligible institutions having only local importance, automatically ensues.

An economist could multiply instances and give them a more precise form. But I am not an economist, and the facts in any case are too well known to need detailed rehearsal. For my purpose is only to indicate the bearing of the development of these corporations upon the change of social life from an individual to a corporate affair. Reactions to the change are psychological, professional, political; they affect the working ideas, beliefs and conduct of all of us.

The sad decline of the farmer cannot be understood except in the light of the industrialization of the country which is coincident with its "corporization." The government is now going to try to do for the collectivizing of the agriculturists the sort of thing that business acumen has already done—temporarily against the desire of the government—for manufactures and transportation. The plight of the uncombined and unintegrated is proof of the extent to which the country is controlled by the corporate idea. Sociologists who concern themselves with rural life are now chiefly occupied with pointing out the influence of urban districts—that is, of those where industrial organization predominates—upon the determination of conditions in country districts.

There are other decays which tell the same story. The old-type artisan, trained by individual apprenticeship for skilled individual work, is disappearing. Mass production by men massed together to operate machines with their minute divisions of labor, is putting him out of business. In many cases, a few weeks at a machine give about all the education—or rather training—that is needed. Mass production causes a kind of mass education in which individual capacity and skill are submerged. While the artisan becomes more of a mechanic and less of an artist, those who are still called artists either put themselves, as writers and designers, at the disposal of organized business, or are pushed out to the edge as eccentric bohemians. The artist remains, one may say, as a surviving individual force, but the esteem in which the calling is socially held in this country measures the degree of his force. The status of the artist in any form of social life affords a fair measure of the state of its culture. The inorganic position of the artist in American life to-day is convincing evidence of what happens to the isolated individual who lives in a society growing corporate.

Attention has recently been called to a new phenomenon in human culture:—the business mind, having its own conversation and language, its own interests, its own intimate groupings in which men of this mind, in their collective capacity, determine the tone of society at large as well as the government of industrial society, and have more political influence than the government itself. I am not concerned here with their political power. The fact significant for present discussion, is that we now have, although without formal or legal status, a mental and moral corporateness for which history affords no parallel. Our indigenous heroes are the Fords and Edisons who typify this mind to the public. Critics may find amusement in ridiculing Rotarians, Kiwanians and Lions, but the latter can well afford to disregard the ridicule because they are representatives of the dominant corporate mentality.

Nowhere is the decline of the old-fashioned individual and individualism more marked than in leisure life, in amusements and sports. Our colleges only follow the movement of the day when they make athletics an organized business, aroused and conducted under paid directors in the spirit of pure collectivism. The formation of theater chains is at once the cause and the effect of the destruction of the older independent life of leisure carried on in separate homes. The radio, the movies, the motor car, all make for a common and aggregate mental and emotional life. With technical exceptions, to be found in special publications and in some portion of all newspapers, the press is the organ of amusement for a hurried leisure time, and it reflects and carries further the formation of mental collectivism by massed methods. Crime, too, is assuming a new form; it is organized and corporate.

Our apartments and our subways are signs of the invasion and decline of privacy. Private "rights" have almost ceased to have a definable meaning. We live exposed to the greatest flood of mass suggestion that any people has ever experienced. The need for united action, and the supposed need of integrated opinion and sentiment, are met by organized propaganda and advertising. The publicity agent is perhaps the most significant symbol of our present social life. There are individuals who resist; but for a time at least, sentiment can be manufactured by mass methods for almost any person or any cause.

These things are not said to be deplored, nor even in order to weigh their merits and demerits. They are merely reported as indications of the nature of our social scene, of the extent to which it is formed and directed by corporate and collective factors toward collective ends. Coincident with these changes in mentality and prestige are basic, if hardly acknowledged, changes in the ideas by which life is interpreted. Industry, again, provides the striking symbols.

What has become of the old-fashioned ideal of thrift? Societies for the promotion of savings among the young were much hurt in their feelings when Henry Ford urged a free scale of expenditures instead of a close scale of personal savings. But his recommendation was in line with all the economic tendencies of the day. Speeded-up mass production demands increased buying. It is promoted by advertising on a vast scale, by instalment selling, by agents skilled in breaking down sales resistance. Hence buying becomes an economic "duty" which is as consonant with the present epoch as thrift was with the period of individualism. For the industrial mechanism depends upon maintaining some kind of an equilibrium between production and consumption. If the equilibrium is disturbed, the whole social structure is affected and prosperity ceases to have a meaning. Replacement and extension of capital are indeed more required than they ever were. But the savings of individuals, as such, are petty and inadequate to the task. New capital is chiefly supplied by the surplus earnings of big corporate organizations, and it becomes meaningless to tell individual buyers that industry can be kept going only by their abstinence from the enjoyments of consumption. The old plea for "sacrifice" loses its force. In effect, the individual is told that by indulging in the enjoyment of free purchasing he performs his economic duty, transferring his surplus income to the corporate store where it can be most effectively used. Virtue departs from mere thrift.

The corresponding change in the ruling conceptions of the older economic theory is, of course, the obligation upon employers to pay high wages. Growing consumption through increased expenditure that effects a still greater amount of production cannot be maintained unless consumers have the wherewithal. The consumption demand of the well-to-do is limited; and their number is limited. Purchase of luxuries by this class has, indeed, become a necessity rather than a vice, since it helps to keep moving the wheels of industry and commerce. Luxury may still be condemned as a vice, just as old habits still show themselves in approving thrift as a virtue. But the condemnation is almost an idle beating of the air, because it goes contrary to the movement of industry and trade. In any case, however, there is a definite limit to the consumption of luxuries, as well as of what used to be called necessary commodities, on the part of the wealthy. The demands that make production and distribution "going concerns" must come from the mass of the people, that is, from workers and those in subordinate salaried positions. Hence the "new economy" based on the idea of the identity of high wages with industrial prosperity.

It is difficult, perhaps impossible, to measure the full import of this revaluation of those concepts of saving and low wages which were basic in the older doctrine. If it merely expressed a change in abstract economic theory, its significance would not be great. But the change in theory is itself a reflex of a social change, which is hardly less than revolutionary. I do not mean that I think that the "new economy" is firmly established as a fact, or that the endless chain of speeding up mass consumption in order to speed up production is either endless or entirely logical. But certain changes do not go backward. Those who have enjoyed high wages and a higher standard of consumption will not be content to return to a lower level. A new condition has been created with which we shall have to reckon constantly in the future. Depressions and slumps will come, but they can never be treated in the future in the casual and fatalistic way in which they have been accepted in the past. They will appear abnormal instead of normal, and society, including the industrial captains, will have to assume a responsibility from which it and they were previously exempt. The gospel of general prosperity in this life will have to meet tests to which that of salvation in the next world, as a compensation for the miseries of this one, was not subjected. "Prosperity" is not such an assured fact in 1930 as it seemed to many to be in the earlier part of 1929. The slump or the depression makes the problem caused by the growth of corporate industry and finance the more acute. An excess income of eight billions a year will only aggravate the economic situation unless it can find outlet in productive channels. It cannot do this unless consumption is sustained. This cannot happen unless organization and control extend from production and distribution to consumption. The alternatives seem to be either a definite expansion of social corporateness to include the average consumer or else economic suffering on a vast scale.

I have said that the instances cited of the reaction of the growing corporateness of society upon social mind and habit were not given in order to be either deplored or approved. They are set forth only to call out the picture of the decline of an individualistic philosophy of life, and the formation of a collectivistic scheme of interdependence, which finds its way into every cranny of life, personal, intellectual, emotional, affecting leisure as well as work, morals as well as economics. But because the purpose was to indicate the decay of the older conceptions, although they are still those that are most loudly and vocally professed, the illustrations given inevitably emphasize those features of growing standardization and mass uniformity which critics justly deplore. It would be unfair, accordingly, to leave the impression that these traits are the whole of the story of the "corporization" of American life.

The things which are criticized are the outward signs of an inner movement toward integration on a scale never known before. "Socialization" is not wholly a eulogistic term nor a desirable process. It involves danger to some precious values; it involves a threat of danger to some things which we should not readily lose. But in spite of much cant which is talked about "service" and "social responsibility," it marks the beginning of a new era of integration. What its ultimate possibilities are, and to what extent these possibilities will be realized, is for the future to tell. The need of the present is to apprehend the fact that, for better or worse, we are living in a corporate age.

It is of the nature of society as of life to contain a balance of opposed forces. Actions and reactions are ultimately equal and counterpart. At present the "socialization" is largely mechanical and quantitative. The system is kept in a kind of precarious balance by the movement toward lawless and reckless overstimulation among individuals. If the chaos and the mechanism are to generate a mind and soul, an integrated personality, it will have to be an intelligence, a sentiment and an individuality of a new type.

Meanwhile, the lawlessness and irregularity (and I have in mind not so much outward criminality as emotional instability and intellectual confusion) and the uniform standardization are two sides of the same emerging corporate society. Hence only in an external sense does society maintain a balance. When the corporateness becomes internal, when, that is, it is realized in thought and purpose, it will become qualitative. In this change, law will be realized not as a rule arbitrarily imposed from without but as the relations which hold individuals together. The balance of the individual and the social will be organic. The emotions will be aroused and satisfied in the course of normal living, not in abrupt deviations to secure the fulfillment which is denied them in a situation which is so incomplete that it cannot be admitted into the affections and yet is so pervasive that it cannot be escaped: a situation which defines an individual divided within himself.

The organism within the context of the environment.

Essential quality for learning — an authentic problem

Try to solve the problem and conceptual part of growth growing as a person

4.   The Lost Individual

The development of a civilization that is outwardly corporate—or rapidly becoming so—has been accompanied by a submergence of the individual. Just how far this is true of the individual's opportunities in action, how far initiative and choice in what an individual does are restricted by the economic forces that make for consolidation, I shall not attempt to say. It is arguable that there has been a diminution of the range of decision and activity for the many along with exaggeration of opportunity of personal expression for the few. It may be contended that no one class in the past has the power now possessed by an industrial oligarchy. On the other hand, it may be held that this power of the few is, with respect to genuine individuality, specious; that those outwardly in control are in reality as much carried by forces external to themselves as are the many; that in fact these forces impel them into a common mold to such an extent that individuality is suppressed.

What is here meant by "the lost individual" is, however, so irrelevant to this question that it is not necessary to decide between the two views. For by it is meant a moral and intellectual fact which is independent of any manifestation of power in action. The significant thing is that the loyalties which once held individuals, which gave them support, direction and unity of outlook on life, have well-nigh disappeared. In consequence, individuals are confused and bewildered. It would be difficult to find in history an epoch as lacking in solid and assured objects of belief and approved ends of action as is the present. Stability of individuality is dependent upon stable objects to which allegiance firmly attaches itself. There are, of course, those who are

[First published as "Individualism, Old and New. II. The Lost Individual," in *New Republic* 61 (5 February 1930): 294–96.]

still militantly fundamentalist in religious and social creed. But their very clamor is evidence that the tide is set against them. For the others, traditional objects of loyalty have become hollow or are openly repudiated, and they drift without sure anchorage. Individuals vibrate between a past that is intellectually too empty to give stability and a present that is too diversely crowded and chaotic to afford balance or direction to ideas and emotion.

Assured and integrated individuality is the product of definite social relationships and publicly acknowledged functions. Judged by this standard, even those who seem to be in control, and to carry the expression of their special individual abilities to a high pitch, are submerged. They may be captains of finance and industry, but until there is some consensus of belief as to the meaning of finance and industry in civilization as a whole, they cannot be captains of their own souls—their beliefs and aims. They exercise leadership surreptitiously and, as it were, absent-mindedly. They lead, but it is under cover of impersonal and socially undirected economic forces. Their reward is found not in what they do, in their social office and function, but in a deflection of social consequences to private gain. They receive the acclaim and command the envy and admiration of the crowd, but the crowd is also composed of private individuals who are equally lost to a sense of social bearings and uses.

The explanation is found in the fact that while the actions promote corporate and collective results, these results are outside their intent and irrelevant to that reward of satisfaction which comes from a sense of social fulfillment. To themselves and to others, their business is private and its outcome is private profit. No complete satisfaction is possible where such a split exists. Hence the absence of a sense of social value is made up for by an exacerbated acceleration of the activities that increase private advantage and power. One cannot look into the inner consciousness of his fellows, but if there is any general degree of inner contentment on the part of those who form our pecuniary oligarchy, the evidence is sadly lacking. As for the many, they are impelled hither and yon by forces beyond their control.

The most marked trait of present life, economically speaking, is insecurity. It is tragic that millions of men desirous of working should be recurrently out of employment; aside from cyclical depressions there is a standing army at all times who have no regular work. We have not any adequate information as to the number of these persons. But the ignorance even as to numbers is slight compared with our inability to grasp the psychological and moral consequences of the precarious condition in which vast multitudes live. Insecurity cuts deeper and extends more widely than bare unemployment. Fear of loss of work, dread of the oncoming of old age, create anxiety and eat into self-respect in a way that impairs personal dignity. Where fears abound, courageous and robust individuality is undermined. The vast development of technological resources that might bring security in its train has actually brought a new mode of insecurity, as mechanization displaces labor. The mergers and consolidations that mark a corporate age are beginning to bring uncertainty into the economic lives of the higher salaried class, and that tendency is only just in its early stage. Realization that honest and industrious pursuit of a calling or business will not guarantee any stable level of life lessens respect for work and stirs large numbers to take a chance of some adventitious way of getting the wealth that will make security possible: witness the orgies of the stock-market in recent days.

The unrest, impatience, irritation and hurry that are so marked in American life are inevitable accompaniments of a situation in which individuals do not find support and contentment in the fact that they are sustaining and sustained members of a social whole. They are evidence, psychologically, of abnormality, and it is as idle to seek for their explanation within the deliberate intent of individuals as it is futile to think that they can be got rid of by hortatory moral appeal. Only an acute maladjustment between individuals and the social conditions under which they live can account for such widespread pathological phenomena. Feverish love of anything as long as it is a change which is distracting, impatience, unsettlement, nervous discontentment, and desire for excitement, are not native to human nature. They are so abnormal as to demand explanation in some deep-seated cause.

I should explain a seeming hypocrisy on the same ground. We are not consciously insincere in our professions of devotion to ideals of "service"; they mean something. Neither the Rotarian nor the big business enterprise uses the term merely as a cloak for "putting something over" which makes for pecuniary gain. But the lady doth protest too much. The wide currency of such professions testifies to a sense of a social function of business which is expressed in words because it is so lacking in fact, and yet which is felt to be rightfully there. If our external combinations in industrial activity were reflected in organic integrations of the desires, purposes and satisfactions of individuals, the verbal protestations would disappear, because social utility would be a matter of course.

Some persons hold that a genuine mental counterpart of the outward social scheme is actually forming. Our prevailing mentality, our "ideology," is said to be that of the "business mind" which has become so deplorably pervasive. Are not the prevailing standards of value those derived from pecuniary success and economic prosperity? Were the answer unqualifiedly in the affirmative, we should have to admit that our outer civilization is attaining an inner culture which corresponds to it, however much we might disesteem the quality of that culture. The objection that such a condition is impossible, since man cannot live by bread, by material prosperity alone, is tempting, but it may be said to beg the question. The conclusive answer is that the business mind is not itself unified. It is divided within itself and must remain so as long as the results of industry as the determining force in life are corporate and collective while its animating motives and compensations are so unmitigatedly private. A unified mind, even of the business type, can come into being only when conscious intent and consummation are in harmony with consequences actually effected. This statement expresses conditions so psychologically assured that it may be termed a law of mental integrity. Proof of the existence of the split is found in the fact that while there is much planning of future development with a view to dividends within large business corporations, there is no corresponding coordinated planning of social development.

The growth of corporateness is arbitrarily restricted. Hence it operates to limit individuality, to put burdens on it, to confuse and submerge it. It crowds more out than it incorporates in an ordered and secure life. It has made rural districts stagnant while bringing excess and restless movement to the city. The restriction of corporateness lies in the fact that it remains on the cash level. Men are brought together on the one side by investment in the same joint stock company, and on the other hand by the fact that the machine compels mass production in order that investors may get their profits. The results affect all society in all its phases. But they are as inorganic as the ultimate human motives that operate are private and egoistic. An economic individualism of motives and aims underlies our present corporate mechanisms, and undoes the individual.

The loss of individuality is conspicuous in the economic region because our civilization is so predominantly a business civilization. But the fact is even more obvious when we turn to the political scene. It would be a waste of words to expatiate on the meaninglessness of present political platforms, parties and issues. The old-time slogans are still reiterated, and to a few these words still seem to have a real meaning. But it is too evident to need argument that on the whole our politics, as far as they are not covertly manipulated in behalf of the pecuniary advantage of groups, are in a state of confusion; issues are improvised from week to week with a constant shift of allegiance. It is impossible for individuals to find themselves politically with surety and efficiency under such conditions. Political apathy broken by recurrent sensations and spasms is the natural outcome.

The lack of secure objects of allegiance, without which individuals are lost, is especially striking in the case of the liberal. The liberalism of the past was characterized by the possession of a definite intellectual creed and program; that was its distinction from conservative parties which needed no formulated outlook beyond defense of things as they were. In contrast, liberals operated on the basis of a thought-out social philosophy, a theory of politics sufficiently definite and coherent to be easily translated into a program of policies to be pursued. Liberalism today is hardly more than a temper of mind, vaguely called forward-looking, but quite uncertain as to where to look and what to look forward to. For many individuals, as well as in its social results, this fact is hardly less than a tragedy. The tragedy may be unconscious for the mass, but they show its reality in their aimless drift, while the more thoughtful are consciously disturbed. For human nature is self-possessed only as it has objects to which it can attach itself.

I do not think it is fantastic to connect our excited and rapacious nationalism with the situation in which corporateness has gone so far as to detach individuals from their old local ties and allegiances but not far enough to give them a new centre and order of life. The most militaristic of nations secures the loyalty of its subjects not by physical force but through the power of ideas and emotions. It cultivates ideals of loyalty, of solidarity, and common devotion to a common cause. Modern industry, technology and commerce have created modern nations in their external form. Armies and navies exist to protect commerce, to make secure the control of raw materials, and to command markets. Men would not sacrifice their lives for the purpose of securing economic gain for a few if the conditions presented themselves to their minds in this bald fashion. But the balked demand for genuine cooperativeness and reciprocal solidarity in daily life finds an outlet in nationalistic sentiment. Men have a pathetic instinct toward the adventure of living and struggling together; if the daily community does not feed this impulse, the romantic imagination pictures a grandiose nation in which all are one. If the simple duties of peace do not establish a common life, the emotions are mobilized in the service of a war that will supply its temporary simulation.

stay
out of
war

I have thus far made no reference to what many persons would consider the most serious and the most overtly evident of all the modes of loss of secure objects of loyalty—religion. It is probably easy to exaggerate the extent of the decadence of religion in an outward sense, church membership, church-going and so on. But it is hardly possible to overstate its decline as a vitally integrative and directive force in men's thought and sentiments. Whether even in the ages of the past that are called religious, religion was itself the actively central force that it is sometimes said to have been may be doubted. But it cannot be doubted that it was the symbol of the existence of conditions and forces that gave unity and a centre to men's views of life. It at least gathered together in weighty and shared symbols a sense of the objects to which men were so attached as to have support and stay in their outlook on life.

Religion does not now effect this result. The divorce of church and state has been followed by that of religion and society. Wherever religion has not become a merely private indulgence, it has become at best a matter of sects and denominations divided from one another by doctrinal differences, and united internally by tenets that have a merely historical origin, and a purely metaphysical or else ritualistic meaning. There is no such bond of social unity as once united Greeks, Romans, Hebrews, and Catholic medieval Europe. There are those who realize what is portended by the loss of religion as an integrating bond. Many of them despair of its recovery through the development of social values to which the imagination and sentiments of individuals can attach themselves with intensity. They wish to reverse the operation and to form the social bond of unity and of allegiance by regeneration of the isolated individual soul.

Aside from the fact that there is no consensus as to what a new religious attitude is to centre itself about, the injunction puts the cart before the horse. Religion is not so much a root of unity as it is its flower or fruit. The very attempt to secure integration for the individual, and through him for society, by means of a deliberate and conscious cultivation of religion, is itself proof of how far the individual has become lost through detachment from acknowledged social values. It is no wonder that when the appeal does not take the form of dogmatic fundamentalism, it tends to terminate in either some form of esoteric occultism or private estheticism. The sense of wholeness which is urged as the essence of religion can be built up and sustained only through membership in a society which has attained a degree of unity. The attempt to cultivate it first in individuals and then extend it to form an organically unified society is fantasy. Indulgence in this fantasy infects such interpretations of American life as are found, to take one signal example, in Waldo Frank's *The Rediscovery of America*.[1] It marks a manner of yearning and not a principle of construction.

For the idea that the outward scene is chaotic because of the machine, which is a principle of chaos, and that it will remain so until individuals reinstitute wholeness within themselves, simply reverses the true state of things. The outward scene, if not fully organized, is relatively so in the corporateness which the machine and its technology have produced; the inner man is the jungle which can be subdued to order only as the forces of organization

---

1. After a brilliant exposition of the dissolution of the European synthesis, he goes on to say "man's need of order and his making of order are his science, his art, his religion; and these are all to be referred to the initial sense of order called the self," quite oblivious of the fact that this doctrine of the primacy of the self is precisely a reaction of the romantic and subjective age to the dissolution he has depicted, having its meaning only in that dissolution.

at work in externals are reflected in corresponding patterns of thought, imagination and emotion. The sick cannot heal themselves by means of their disease, and disintegrated individuals can achieve unity only as the dominant energies of community life are incorporated to form their minds. If these energies were, in reality, mere strivings for private pecuniary gain, the case would indeed be hopeless. But they are constituted by a collective art of technology, which individuals merely deflect to their private ends. There are the beginnings of an objective order through which individuals may get their bearings.

Conspicuous signs of the disintegration of individuality due to failure to reconstruct the self so as to meet the realities of present social life have not been mentioned. In a census that was taken among leaders of opinion concerning the urgency of present social problems, the state of law, the courts, lawlessness and criminality stood at the head of the list, and by a considerable distance. We are even more emphatically than when Kipling wrote the words, the people that make "the laws they flout, and flout the laws they make." We combine an ardor unparalleled in history for "passing" laws with a casual and deliberate disregard for them when they are on the statute books. We believe—to judge by our legislative actions—that we can create morals by law (witness the prohibition amendment for an instance on a large scale) and neglect the fact that all laws except those which regulate technical procedures are registrations of existing social customs and their attendant moral habits and purposes. I can, however, only think of this phenomenon as a symptom, not as a cause. It is a natural expression of a period in which changes in the structure of society have dissolved old bonds and allegiances. We attempt to make good this social relaxation and dissolution by legal enactments, while the actual disintegration discloses itself in the lawlessness which reveals the artificial character of this method of securing social integrity.

Volumes could be formed by collecting articles and editorials written about relaxation of traditional moral codes. A movement has caught public attention, which, having for some obscure reason assumed the name "humanism," proposes restraint and moderation, exercised in and by the higher volition of individuals, as the solution of our ills. It finds that naturalism as practiced by artists and mechanism as taught by philosophers who take their clew from natural science, have broken down the inner laws and imperatives which can alone bring order and loyalty. I should be glad to be able to believe that artists and intellectuals have any such power in their hands; if they had, after using it to bring evil to society, they might change face and bring healing to it. But a sense of fact, together with a sense of humor, forbids the acceptance of any such belief. Literary persons and academic thinkers are now, more than ever, effects, not causes. They reflect and voice the disintegration which new modes of living, produced by new forms of industry and commerce, have introduced. They give witness to the unreality that has overtaken traditional codes in the face of the impact of new forces; indirectly, they proclaim the need of some new synthesis. But this synthesis can be humanistic only as the new conditions are themselves taken into account and are converted into the instrumentalities of a free and humane life. I see no way to "restrain" or turn back the industrial revolution and its consequences. In the absence of such a restraint (which would be efficacious if only it could occur), the urging of some inner restraint through the exercise of the higher personal will, whatever that may be, is itself only a futile echo of just the old individualism that has so completely broken down.

There are many phases of life which illustrate to anyone who chooses to think in terms of realities instead of words the utter irrelevance of the proposed remedy to actual conditions. One might take the present estate of amusements, of the movies, the radio, and organized vicarious sport, and ask just how this powerful eruption in which the resources of technology are employed for economic profit is to be met by the application of the inner *frein* or brake. Perhaps the most striking instance is found in the disintegration due to changes in family life and sex morale. It was not deliberate human intention that undermined the traditional household as the centre of industry and education and as the focus of moral training; that sapped the older institution of enduring marriage. To ask the individuals who suffer the consequences of the general undermining and sapping to put an end to the consequences by acts of personal volition is merely to profess faith in moral magic. Recovery of individuals capable of stable and effective self-control can be had only as there is first a humbler exercise of will to observe existing social realities and to direct them according to their own potentialities.

Instances of the flux in which individuals are loosened from the ties that once gave order and support to their lives are glaring. They are indeed so glaring that they blind our eyes to the causes which produce them. Individuals are groping their way through situations which they do not direct and which do not give them direction. The beliefs and ideals that are uppermost in their consciousness are not relevant to the society in which they outwardly act and which constantly reacts upon them. Their conscious ideas and standards are inherited from an age that has passed away; their minds, as far as consciously entertained principles and methods of interpretation are concerned, are at odds with actual conditions. This profound split is the cause of distraction and bewilderment.

Individuals will refind themselves only as their ideas and ideals are brought into harmony with the realities of the age in which they act. The task of attaining this harmony is not an easy one. But it is more negative than it seems. If we could inhibit the principles and standards that are merely traditional, if we could slough off the opinions that have no living relationship to the situations in which we live, the unavowed forces that now work upon us unconsciously but unremittingly would have a chance to build minds after their own pattern, and individuals might, in consequence, find themselves in possession of objects to which imagination and emotion would stably attach themselves.

I do not mean, however, that the process of rebuilding can go on automatically. Discrimination is required in order to detect the beliefs and institutions that dominate merely because of custom and inertia, and in order to discover the moving realities of the present. Intelligence must distinguish, for example, the tendencies of the technology which produce the new corporateness from those inheritances proceeding out of the individualism of an earlier epoch which arrest and divide the operation of the new dynamics. It is difficult for us to conceive of individualism except in terms of stereotypes derived from former centuries. Individualism has been identified with ideas of initiative and invention that are bound up with private and exclusive economic gain. As long as this conception possesses our minds, the ideal of harmonizing our thought and desire with the realities of present social conditions will be interpreted to mean accommodation and surrender. It will even be understood to signify rationalization of the evils of existing society. A stable recovery of individuality waits upon an elimination of the older economic and political individualism, an elimination which will liberate imagination and endeavor for the task of making corporate society contribute to the free culture of its members. Only by economic revision can the sound element in the older individualism—equality of opportunity—be made a reality.

It is the part of wisdom to note the double meaning of such ideas as "acceptance." There is an acceptance that is of the intellect; it signifies facing facts for what they are. There is another acceptance that is of the emotions and will; that involves commitment of desire and effort. So far are the two from being identical that acceptance in the first sense is the precondition of all intelligent refusal of acceptance in the second sense. There is a prophetic aspect to all observation; we can perceive the meaning of what exists only as we forecast the consequences it entails. When a situation is as confused and divided within itself as is the present social estate, choice is implicated in observation. As one perceives different tendencies and different possible consequences, preference inevitably goes out to one or the other. Because acknowledgment in thought brings with it intelligent discrimination and choice, it is the first step out of confusion, the first step in forming those objects of significant allegiance out of which stable and efficacious individuality may grow. It might even perform the miracle of rendering conservatism relevant and thoughtful. It certainly is the prerequisite of an anchored liberalism.

5. Toward a New Individualism

Our material culture, as anthropologists would call it, is verging upon the collective and corporate. Our moral culture, along with our ideology, is, on the other hand, still saturated with ideals and values of an individualism derived from the pre-scientific, pre-technological age. Its spiritual roots are found in medieval religion, which asserted the ultimate nature of the individual soul and centered the drama of life about the destiny of that soul. Its institutional and legal concepts were framed in the feudal period.

This moral and philosophical individualism anteceded the rise of modern industry and the era of the machine. It was the context in which the latter operated. The apparent subordination of the individual to established institutions often conceals from recognition the vital existence of a deep-seated individualism. But the fact that the controlling institution was the Church should remind us that in ultimate intent it existed to secure the salvation of the individual. That this individual was conceived as a soul, and that the end served by the institution was deferred to another and everlasting life conceal from contemporary realization the underlying individualism. In its own time, its substance consisted in just this eternal spiritual character of the personal soul; the power of the established institutions proceeded from their being the necessary means of accomplishing the supreme end of the individual.

The early phase of the industrial revolution wrought a great transformation. It gave a secular and worldly turn to the career of the individual, and it liquefied the static property concepts of feudalism by the shift of emphasis from agriculture to manufac-

[First published as "Toward a New Individualism. The Third Article in Professor Dewey's Series, 'Individualism, Old and New,'" in *New Republic* 62 (19 February 1930): 13–16.]

turing. Still, the idea persisted that property and reward were intrinsically individual. There were, it is true, incompatible elements in the earlier and later versions of individualism. But a fusion of individual capitalism, of natural rights, and of morals founded in strictly individual traits and values remained, under the influence of Protestantism, the dominant intellectual synthesis.

The basis of this synthesis was destroyed, however, by the later development of the industrial system, which brought about the merging of personal capacity, effort and work into collective wholes. Meanwhile, the control of natural energies eliminated time and distance, so that action once adapted to local conditions was swallowed up in complex undertakings of indefinite extent. Yet the older mental equipment remained after its causes and foundations had disappeared. This, fundamentally, is the inner division out of which spring our present confusion and insincerities.

The earlier economic individualism had a definite creed and function. It sought to release from legal restrictions man's wants and his efforts to satisfy those wants. It believed that such emancipation would stimulate latent energy into action, would automatically assign individual ability to the work for which it was suited, would cause it to perform that work under stimulus of the advantage to be gained, and would secure for capacity and enterprise the reward and position to which they were entitled. At the same time, individual energy and savings would be serving the needs of others, and thus promoting the general welfare and effecting a general harmony of interests.

We have gone a long way since this philosophy was formulated. Today, the most stalwart defenders of this type of individualism do not venture to repeat its optimistic assertions. At most, they are content to proclaim its consistency with unchanging human nature—which is said to be moved to effort only by the hope of personal gain—and to paint dire pictures of the inevitable consequences of change to any other regime. They ascribe all the material benefits of our present civilization to this individualism—as if machines were made by the desire for money profit, not by impersonal science; and as if they were driven by money alone, and not by electricity and steam under the direction of a collective technology.

In America, the older individualism assumed a romantic form. It was hardly necessary to elaborate a theory which equated personal gain with social advance. The demands of the practical situation called for the initiative, enterprise and vigor of individuals in all immediate work that urgently asked for doing, and their operation furthered the national life. The spirit of the time is expressed by Dr. Crothers, whose words Mr. Sims has appropriately taken for part of the text of his "Adventurous America":

> If you would understand the driving power of America, you must understand "the divers discontented and impatient young men" who in each generation have found an outlet for their energy. . . . The noises which disturb you are not the cries of an angry proletariat, but are the shouts of eager young people who are finding new opportunities. . . . They represent today the enthusiasm of a new generation. They represent the Oregons and Californias toward which sturdy pioneers are moving undisturbed by obstacles. This is what the social unrest means in America.

If that is not an echo of the echo of a voice of long ago, I do not know what it is. I do not, indeed, hear the noises of an angry proletariat; but I should suppose the sounds heard are the murmurs of lost opportunities, along with the din of machinery, motor cars and speakeasies, by which the murmurs of discontent are drowned, rather than shouts of eagerness for adventurous opportunity.

The European version of the older individualism had its value and temporal justification because the new technology needed liberation from vexatious legal restrictions. Machine industry was itself in a pioneer condition, and those who carried it forward against obstacles of lethargy, skepticism and political obstruction were deserving of special reward. Moreover, accumulation of capital was thought of in terms of enterprises that today would be petty; there was no dream of the time when it would reach such a mass that it would determine the legal and political order. Poverty had previously been accepted as a dispensation of nature that was inevitable. The new industry promised a way out, at least to those possessed of energy and will to save and accumulate. But there was no anticipation of a time when the development of machine technology would afford the material basis for reasonable ease and comfort and of extensive leisure for all.

The shift that makes the older individualism a dying echo is more marked as well as more rapid in this country. Where is the wilderness which now beckons creative energy and affords untold opportunity to initiative and vigor? Where is the pioneer who goes forth rejoicing, even in the midst of privation, to its conquest? The wilderness exists in the movie and the novel, and the children of the pioneers, who live in the midst of surroundings artificially made over by the machine, enjoy pioneer life idly in the vicarious film. I see little social unrest which is the straining of energy for outlet in action; I find rather the protest against a weakening of vigor and a sapping of energy that emanate from the absence of constructive opportunity; and I see a confusion that is an expression of the inability to find a secure and morally rewarding place in a troubled and tangled economic scene.

Because of the bankruptcy of the older individualism, those who are aware of the break-down often speak and argue as if individualism were itself done and over with. I do not suppose that those who regard socialism and individualism as antithetical really mean that individuality is going to die out or that it is not something intrinsically precious. But in speaking as if the only individualism were the local episode of the last two centuries, they play into the hands of those who would keep it alive in order to serve their own ends, and they slur over the chief problem—that of remaking society to serve the growth of a new type of individual. There are many who believe that socialism of some form is needed to realize individual initiative and security on a wide scale. They are concerned about the restriction of power and freedom to a few in the present regime, and they think that collective social control is necessary, at least for a time, in order to achieve its advantages for all. But they too often seem to assume that the result will be merely an extension of the earlier individualism to the many.

Such thinking treats individualism as if it were something static, having a uniform content. It ignores the fact that the mental and moral structure of individuals, the pattern of their desires and purposes, change with every great change in social constitution. Individuals who are not bound together in associations, whether domestic, economic, religious, political, artistic or educational, are monstrosities. It is absurd to suppose that the ties which hold them together are merely external and do not react into mentality and character, producing the framework of personal disposition.

The tragedy of the "lost individual" is due to the fact that while individuals are now caught up into a vast complex of associations, there is no harmonious and coherent reflection of the import of these connections into the imaginative and emotional outlook on life. This fact is of course due in turn to the absence of harmony within the state of society. There is an undoubted circle. But it is a vicious circle only as far as men decline to accept—in the intellectual, observing and inquiring spirit defined in the previous chapter—the realities of the social estate, and because of this refusal either surrender to the division or seek to save their individuality by escape or sheer emotional revolt. The habit of opposing the corporate and collective to the individual tends to the persistent continuation of the confusion and uncertainty. It distracts attention from the crucial issue: How shall the individual refind himself in an unprecedentedly new social situation, and what qualities will the new individualism exhibit?

That the problem is not merely one of extending to all individuals the traits of economic initiative, opportunity and enterprise; that it is one of forming a new psychological and moral type, is suggested by the great pressure now brought to bear to effect conformity and standardization of American opinion. Why should regimentation, the erection of an average struck from the opinions of large masses into regulative norms, and in general the domination of quantity over quality, be so characteristic of present American life? I see but one fundamental explanation. The individual cannot remain intellectually a vacuum. If his ideas and beliefs are not the spontaneous function of a communal life in which he shares, a seeming consensus will be secured as a substitute by artificial and mechanical means. In the absence of mentality that is congruous with the new social corporateness that is coming into being, there is a desperate effort to fill the void by external agencies which obtain a factitious agreement.

In consequence, our uniformity of thought is much more superficial than it seems to be. The standardization is deplorable, but one might almost say that one of the reasons it is deplorable is because it does not go deep. It goes far enough to effect suppression of original quality of thought, but not far enough to achieve enduring unity. Its superficial character is evident in its instability. All agreement of thought obtained by external means, by repression and intimidation, however subtle, and by calculated propaganda and publicity, is of necessity superficial; and whatever is superficial is in continual flux. The methods employed produce mass credulity, and this jumps from one thing to another according to the dominant suggestions of the day. We think and feel alike—but only for a month or a season. Then comes some other sensational event or personage to exercise a hypnotizing uniformity of response. At a given time, taken in cross-section, conformity is the rule. In a time span, taken longitudinally, instability and flux dominate. . . . I suppose there are others who have a feeling of irritation at such terms as "radioconscious" and "air-minded," now so frequently forced upon us. I do not think the irritation is wholly due to linguistic causes. It testifies to a half-conscious sense of the external ways in which our minds are formed and swayed and of the superficiality and inconsistency of the result.

There are, I suppose, those who fancy that the emphasis which I put upon the corporateness of existing society in the United States is in effect, even if not in the writer's conscious intent, a plea for greater conformity than now exists. Nothing could be further from the truth. Identification of society with a level, whatever it be, high as well as low, of uniformity is just another evidence of that distraction because of which the individual is lost. Society is of course but the relations of individuals to one another in this form and that. And all relations are interactions, not fixed molds. The particular interactions that compose a human society include the give and take of participation, of a sharing that increases, that expands and deepens, the capacity and significance of the interacting factors. Conformity is a name for the absence of vital interplay; the arrest and benumbing of communication. As I have been trying to say, it is the artificial substitute used to hold men together in lack of associations that are incorporated into inner dispositions of thought and desire. I often wonder what meaning is given to the term "society" by those who oppose it to the intimacies of personal intercourse, such as those of friendship. Presumably they have in their minds a picture of rigid institutions or some set and external organization. But an institution that is other than the structure of human contact and intercourse is a fossil of some past society; organization, as in any living organism, is the cooperative consensus of multitudes of cells, each living in exchange with others.

I should suppose that the more intelligent of those who wield the publicity agencies which produce conformity would be disturbed at beholding their own success. I can easily understand that they should have a cynical sense of their ability to obtain the results they want at a given time; but I should think they would fear that like-mindedness might, at a critical juncture, veer in an unexpected direction and turn with equal unanimity against the things and interests it has been manipulated to support. Crowd psychology is dangerous in its instability. To rely upon it for permanent support is playing with a fire that may get out of control. Conformity is enduringly effective when it is a spontaneous and largely unconscious manifestation of the agreements that spring from genuine communal life. An artificially induced uniformity of thought and sentiment is a symptom of an inner void. Not all of it that now exists is intentionally produced; it is not the result of deliberate manipulation. But it is, on the other hand, the result of causes so external as to be accidental and precarious.

The "joining" habit of the average American, and his excessive sociability, may well have an explanation like that of conformity. They, too, testify to nature's abhorrence of that vacuum which the passing of the older individualism has produced. We should not be so averse to solitude if we had, when we were alone, the companionship of communal thought built into our mental habits. In the absence of this communion, there is the need for reinforcement by external contact. Our sociability is largely an effort to find substitutes for that normal consciousness of connection and union that proceeds from being a sustained and sustaining member of a social whole.

Just as the new individualism cannot be achieved by extending the benefits of the older economic individualism to more persons, so it cannot be obtained by a further development of generosity, good will and altruism. Such traits are desirable, but they are also more or less constant expressions of human nature. There is much in the present situation that stimulates them to active operation. They are probably more marked features of American life than of that of any other civilization at any time.

Our charity and philanthropy are partly the manifestation of an uneasy conscience. As such a manifestation, they testify to a realization that a régime of industry carried on for private gain does not satisfy the full human nature of even those who profit by it. The impulse and need which the existing economic régime chokes, through preventing its articulated expression, find outlet in actions that acknowledge a social responsibility which the system as a system denies. Hence the development of philanthropic measures is not only compensatory to a stifling of human nature undergone in business, but it is in a way prophetic. Construction is better than relief; prevention than cure. Activities by way of relief of poverty and its attendant mental strains and physical ills—and our philanthropic activities including even the endowment of educational institutions have their ultimate causes in the existence of economic insecurity and distress—suggest, in dim forecast, a society in which daily occupations and relationships will give independence and substantial living to all normal individuals who share in its ongoings, reserving relief for extraordinary emergencies. One does not need to reflect upon the personal motives of great philanthropists to see in what they do an emphatic record of the breakdown of our existing economic organization.

For the chief obstacle to the creation of a type of individual whose pattern of thought and desire is enduringly marked by consensus with others, and in whom sociability is one with cooperation in all regular human associations, is the persistence of that feature of the earlier individualism which defines industry and commerce by ideas of private pecuniary profit. Why, once more, is there such zeal for standardized likeness? It is not, I imagine, because conformity for its own sake appears to be a great boon. It is rather because a certain kind of conformity gives defense and protection to the pecuniary features of our present regime. The foreground may be filled with depiction of the horror of change, and with clamor for law and order and the support of the Constitution. But behind there is desire for perpetuation of that regime which defines individual initiative and ability by success in conducting business so as to make money.

It is not too much to say that the whole significance of the older individualism has now shrunk to a pecuniary scale and measure. The virtues that are supposed to attend rugged individualism may be vocally proclaimed, but it takes no great insight to see that what is cherished is measured by its connection with those activities that make for success in business conducted for personal gain. Hence, the irony of the gospel of "individualism" in business conjoined with suppression of individuality in thought and speech. One cannot imagine a bitterer comment on any professed individualism than that it subordinates the only creative individuality—that of mind—to the maintenance of a regime which gives the few an opportunity for being shrewd in the management of monetary business.

It is claimed, of course, that the individualism of economic self-seeking, even if it has not produced the adjustment of ability and reward and the harmony of interests earlier predicted, has given us the advantage of material prosperity. It is not needful to raise here the question of how far that material prosperity extends. For it is not true that its moving cause is pecuniary individualism. That has been the cause of some great fortunes, but not of national wealth; it counts in the process of distribution, but not in ultimate creation. Scientific insight taking effect in machine technology has been the great productive force. For the most part, economic individualism interpreted as energy and enterprise devoted to private profit, has been an adjunct, often a parasitical one, to the movement of technical and scientific forces.

The scene in which individuality is created has been transformed. The pioneer, such as is depicted in the quotation from Crothers, had no great need for any ideas beyond those that sprang up in the immediate tasks in which he was engaged. His intellectual problems grew out of struggle with the forces of physical nature. The wilderness was a reality and it had to be subdued. The type of character that evolved was strong and hardy, often picturesque, and sometimes heroic. Individuality was a reality because it corresponded to conditions. Irrelevant traditional ideas in religion and morals were carried along, but they were reduced to a size where they did no harm; indeed, they could easily be interpreted in such a way as to be a reinforcement to the sturdy and a consolation to the weak and failing.

But it is no longer a physical wilderness that has to be wrestled with. Our problems grow out of social conditions: they concern human relations rather than man's direct relationship to physical nature. The adventure of the individual, if there is to be any venturing of individuality and not a relapse into the deadness of complacency or of despairing discontent, is an unsubdued social frontier. The issues cannot be met with ideas improvised for the occasion. The problems to be solved are general, not local. They concern complex forces that are at work throughout the whole country, not those limited to an immediate and almost face-to-face environment. Traditional ideas are more than irrelevant. They are an encumbrance; they are the chief obstacle to the formation of a new individuality integrated within itself and with a liberated function in the society wherein it exists. A new individualism can be achieved only through the controlled use of all the resources of the science and technology that have mastered the physical forces of nature.

They are not controlled now in any fundamental sense. Rather do they control us. They are indeed physically controlled. Every factory, power-house and railway system testifies to the fact that we have attained this measure of control. But control of power through the machine is not control of the machine itself. Control of the energies of nature by science is not controlled use of science. We are not even approaching a climax of control; we are hardly at its feeble beginnings. For control is relative to consequences, ends, values; and we do not manage, we hardly have commenced to dream of managing, physical power for the sake of projected purposes and prospective goods. The machine took us unawares and unprepared. Instead of forming new purposes commensurate with its potentialities, we accordingly tried to make it the servant of aims that were the expression of an age when mastery of natural energies on any large scale was the fantasy of magic. As Clarence Ayres has said: "Our industrial revolution began, as some historians say, with half a dozen technical improvements in the textile industry; and it took us a century to realize that anything of moment had happened to us beyond the obvious improvement of spinning and weaving."

I do not say that the aims and values of the earlier day were petty in themselves. But they are almost inconceivably petty in comparison with the means now at our command—if we had an imagination large enough to encompass their potential uses. They are worse than petty; they are confusing and distracting when men are confronted with the physical instrumentalities and agencies which, in the lack of comprehensive purpose and concerted planning, work blindly and carry us drifting hither and yon. I cannot obtain intellectual, moral or esthetic satisfaction from the professed philosophy which animates Bolshevik Russia. But I am sure that the future historian of our times will combine admiration of those who had the imagination first to see that the resources of technology might be directed by organized planning to serve chosen ends with astonishment at the intellectual and moral hebetude of other peoples who were technically so much further advanced.

There is no greater sign of the paralysis of the imagination which custom and involvement in immediate detail can induce than the belief, sedulously propagated by some who pride themselves on superior taste, that the machine is itself the source of our troubles. Of course immense potential resources impose responsibility, and it has yet to be demonstrated whether human capacity can rise to utilization of the opportunities which the machine and technology have opened to us. But it is hard to think of anything more childish than the animism that puts the blame on machinery. For machinery means an undreamed-of reservoir of power. If we have harnessed this power to the dollar rather than to the liberation and enrichment of human life, it is because we have been content to stay within the bounds of traditional aims and values although we are in possession of a revolutionary transforming instrument. Repetition of the older credo of individualism is but the evidence of contentment within these bonds. I for one think it is incredible that this particular form of confession of inferiority will endure very much longer. When we begin to ask what can be done with the machine for the creation and fulfillment of values corresponding to its potency and begin organized planning to effect these goods, a new individual correlative to the realities of the age in which we live will also begin to take form.

Revolt against the machine as the author of social evils usually has an esthetic origin. A more intellectual and quasi-philosophic reaction finds natural science to be their source; or if not science itself (which is allowed to be all very well if it keeps its appropriate humble place) then the attitude of those who depend upon science as an organ of vision and light. Contempt for nature is understandable, at least historically; even though it seems both intellectually petty and morally ungracious to feel contempt for the matrix of our being and the inescapable condition of our lives. But that men should fear and dislike the method of approach to nature I do not find understandable. The eye sees many foul things and the arm and hand do many cruel things. Yet the fanatic who would pluck out the eye and cut off the arm is recognized for what he is. Science, one may say, is but the extension of our natural organs of approach to nature. And I do not mean merely an extension in quantitative range and penetration, as a microscope multiplies the capacity of the unaided eye, but an extension of insight and understanding through bringing relationships and interactions into view. Since we must in any case approach nature in some fashion and by some path—if only that of death—I confess my total inability to understand those who object to an intelligently controlled approach—for that is what science is.

The only way in which I can obtain any sympathetic realization of their attitude is by recalling that there have been those who have professed adoration of science—writing it with a capital S—; those who have thought of it not as a method of approach but as a kind of self-enclosed entity and end in itself, a new theology of self-sufficient authoritatively revealed inherent and absolute Truth. It would, however, seem simpler to correct their misapprehension than first to share it and then to reverse their worship into condemnation. The opposite of intelligent method is no method at all or blind and stupid method. It is a curious state of mind which finds pleasure in setting forth the "limits of science." For the intrinsic limit of knowledge is simply ignorance; and the point in extolling ignorance is not clear except when expressed by those who profit by keeping others in ignorance. There is of course an extrinsic limit of science. But that limitation lies in the ineptitude of those who put it to use; its removal lies in rectification of its use, not in abuse of the thing used.

This reference to science and technology is relevant because they are the forces of present life which are finally significant. It is through employing them with understanding of their possible import that a new individualism, consonant with the realities of the present age, may be brought into operative being. There are many levels and many elements in both the individual and his relations. Neither can be comprehended nor dealt with in mass. Discriminative sensitivity, selection, is imperative. Art is the fruit of such selection when it is given objective effect. The art which our times needs in order to create a new type of individuality is the art which, being sensitive to the technology and science that are the moving forces of our time, will envisage the expansive, the social, culture which they may be made to serve. I am not anxious to depict the form which this emergent individualism will assume. Indeed, I do not see how it can be described until more progress has been made in its production. But such progress will not be initiated until we cease opposing the socially corporate to the individual, and until we develop a constructively imaginative observation of the role of science and technology in actual society. The greatest obstacle to that vision is, I repeat, the perpetuation of the older individualism now reduced, as I have said, to the utilization of science and technology for ends of private pecuniary gain. I sometimes wonder if those who are conscious of present ills but who direct their blows of criticism at everything except this obstacle are not stirred by motives which they unconsciously prefer to keep below consciousness.

6. Capitalistic or Public Socialism?

I once heard a distinguished lawyer say that the earlier American ideas about individual initiative and enterprise could be recovered by an amendment of a few lines to the federal Constitution. The amendment would prohibit all joint stock enterprises and permit only individual liability to have a legal status. He was, I think, the only unadulterated Jeffersonian Democrat I have ever met. He was also logical. He did not delude himself into supposing that the pioneer gospel of personal initiative, enterprise, energy and reward could be maintained in an era of aggregated corporate capital, of mass production and distribution, of impersonal ownership and of ownership divorced from management. Our political life, however, continues to ignore the change that has taken place except as circumstances force it to take account of it in sporadic matters.

The myth is still current that socialism desires to use political means in order to divide wealth equally among all individuals, and that it is consequently opposed to the development of trusts, mergers and consolidated business in general. It is regarded, in other words, as a kind of arithmetically fractionized individualism. This notion of socialism is of the sort that would naturally be entertained by those who cannot get away from the inherent conception of the individual as an isolated and independent unit. In reality, Karl Marx was the prophet of just this period of economic consolidation. If his ghost hovers above the American scene, it must find legitimate satisfaction in our fulfilment of his predictions.

In these predictions, however, Marx reasoned too much from psychological economic premises and depended too little upon

[First published as "Capitalistic or Public Socialism? The Fourth Article in Professor Dewey's Series, 'Individualism, Old and New,'" in *New Republic* 62 (5 March 1930): 64–67.]

technological causes—the application of science to steam, electricity and chemical processes. That is to say, he argued to an undue extent from an alleged constant appropriation by capitalists of all surplus values created by the workers—surplus being defined as anything above the minimum needed for their continued subsistence. He had no conception, moreover, of the capacity of expanding industry to develop new inventions so as to develop new wants, new forms of wealth, new occupations; nor did he imagine that the intellectual ability of the employing class would be equal to seeing the need for sustaining consuming power by high wages in order to keep up production and its profits. This explains why his prediction of a revolution in political control, caused by the general misery of the masses and resulting in the establishment of a socialistic society, has not been realized in this country. Nevertheless, the issue which he raised— the relation of the economic structure to political operations—is one that actively persists.

Indeed, it forms the only basis of present political questions. An intelligent and experienced observer of affairs at Washington has said that all political questions which he has heard discussed in Washington come back ultimately to problems connected with the distribution of income. Wealth, property and the processes of manufacturing and distribution—down to retail trade through the chain system—can hardly be socialized in outward effect without a political repercussion. It constitutes an ultimate issue which must be faced by new or existing political parties. There is still enough vitality in the older individualism to offer a very serious handicap to any party or program which calls itself by the name of Socialism. But in the long run, the realities of the situation will exercise control over the connotations which, for historical reasons, cling to a word. In view of this fact, the fortunes of a party called by a given name are insignificant.

In one important sense, the fundamental character of the economic question is not ignored in present politics. The dominant party has officially constituted itself the guardian of prosperity; it has gone further and offered itself as the author of prosperity. It has insinuated itself in that guise into the imagination of a sufficient number of citizens and voters so that it owes its continuing domination to its identification with prosperity. Our presidential elections are upon the whole determined by fear. Hundreds of thousands of citizens who vote independently or for Democratic candidates at local elections and in off-year congressional elections regularly vote the Republican ticket every four years. They do so because of a vague but influential dread lest a monkey-wrench be thrown into the economic and financial machine. The dread is as general among the workers as among small traders and storekeepers. It is basically the asset that keeps the dominant party in office. Our whole industrial scheme is so complex, so delicately interdependent in its varied parts, so responsive to a multitude of subtle influences, that it seems definitely better to the mass of voters to endure the ills they may already suffer rather than take the chance of disturbing industry. Even in the election of 1928, in spite of both the liquor and the Catholic issue, this was, I believe, the determining factor.

Moreover, the fact that Hoover offered himself to the popular imagination as a man possessed of the engineer's rather than the politician's mind was a great force. Engineering has accomplished great things; its triumphs are everywhere in evidence. The miracles that it has wrought have given it the prestige of magical wonder-working. A people sick of politicians felt in some half-conscious way that the mind, experience and gifts of an engineer would bring healing and order into our political life. It is impossible to present statistics as to the exact force of the factors mentioned. Judgment on the two points, especially the latter, must remain a matter of opinion. But the identification of the Republican party with the maintenance of prosperity cannot be denied, and the desire for the engineer in politics is general enough to be at least symptomatic.

Prosperity is largely a state of mind, and belief in it is even more so. It follows that skepticism about its extent is of little importance when the mental tide runs with the idea. Although figures can be quoted to show how spotty it is, and how inequitably its economic conditions are distributed, they are all to no avail. What difference does it make that eleven thousand people, having each an annual income of over $100,000, appropriated in 1927 about one twenty-fifth of the net national income? What good does it do to cite official figures showing that only 20 percent of the income of the favored eleven thousand came from salaries and from profits of the businesses they were personally engaged in, while the remaining 80 percent was derived from investments, speculative profits, rent, etc.? That the total earnings of eight million wage workers should be only four times the amount of what the income-tax returns frankly call the "unearned" income of the eleven thousand millionaires goes almost without notice. Moreover, income from investments in corporate aggregations increases at the expense of that coming from enterprises personally managed. For anyone to call attention to this discrepancy is considered an aspersion on our rugged individualism and an attempt to stir up class feeling. Meanwhile, the income-tax returns for 1928 show that in seven years the number of persons having an annual income of more than $1,000,000 has increased from sixty-seven to almost five hundred, twenty-four of whom had incomes of over $10,000,000 each.

Nevertheless, the assumption of guardianship of prosperity by a political party means the assumption of responsibility, and in the long run the ruling economic-political combination will be held to account. The over-lords will have to do something to make good. This fact seems to me to be the centre of the future political situation. Discussion of the prospective political development in connection with corporate industry may at least start from the fact that the industries which used to be regarded as staple, as the foundations of sound economy, are depressed. The plight of agriculture, of the coal and textile industries, is well known. The era of great railway expansion has come to a close; the building trades have a fluctuating career. The counterpart of this fact is that the now flourishing industries are those connected with and derived from new technical developments. Without the rapid growth in the manufacture and sale of automobiles, radios, airplanes, etc.; without the rapid development of new uses for electricity and super-power, prosperity in the last few years would hardly have been even a state of mind. Economic stimulus has come largely from these new uses for capital and labor; surplus funds drawn from them have kept the stock market and other forms of business actively going. At the same time, these newer developments have accelerated the accumulation and concentration of super-fortunes.

These facts seem to suggest the issue of future politics. The fact of depression has already influenced political action in legislation and administration. What will happen when industries now new become in turn overcapitalized and consumption does not keep up in proportion to investment in them; when they, too, have an excess capacity of production? There are now, it is estimated, eight billions of surplus savings a year, and the amount is increasing. Where is this capital to find its outlet? Diversion into the stock market gives temporary relief, but the resulting inflation is a "cure" which creates a new disease. If it goes into the expansion of industrial plants, how long will it be before they, too, "overproduce"? The future seems to hold in store an extension of political control in the social interest. We already have the Interstate Commerce Commission, the Federal Reserve Board, and now the Farm Relief Board—a socialistic undertaking on a large scale sponsored by the party of individualism. The probabilities seem to favor the creation of more such boards in the future, in spite of all concomitant denunciations of bureaucracy and proclamations that individualism is the source of our national prosperity.

The tariff question, too, is undergoing a change. Now, it is the older industries which, being depressed, clamor for relief. The "infant" industries are those which are indifferent, and which, with their growing interest in export trade, are likely to become increasingly indifferent or hostile. The alignment of political parties has not indeed been affected so far by economic changes—beyond the formation of insurgent blocs within the old parties. But this fact only conceals from view the greater fact that, under cover of the old parties, legislation and administration have taken on new functions due to the impact of trade and finance. The most striking example, of course, is the effort to use governmental agencies and large public funds to put agriculture on a parity with other forms of industry. The case is the more significant because the farmers form the part of the population that has remained most faithful to the old individualistic philosophy, and because the movement is definitely directed to bringing them within the scope of collective and corporate action. The policy of using public works to alleviate unemployment in times of depression is another, if lesser, sign of the direction which political action is taking.

The question of whether and how far the newer industries will follow the cycle of the older and now depressed ones, becoming overcapitalized, overproductive in capacity and overcharged with carrying costs, is, of course, a speculative one. The negative side of the argument demands, however, considerable optimism. It is at least reasonably certain that if depression sets in with them, the process of public intervention and public control will be repeated. And in any case, nothing can permanently exclude political action with reference to old age and unemployment. The scandalous absence even of public inquiry and statistics is emphasized at present by the displacement of workers through technical developments, and by the lowering, because of speeded-up processes, of the age-limit at which workers can be profitably employed. Unemployment, on the scale at which it now "normally" exists—to say nothing of its extent during cyclic periods of depression—is a confession of the breakdown of unregulated individualistic industry conducted for private profit. Coal miners and even farmers may go unheeded, but not so the industrial city workers. One of the first signs of the reawakening of an aggressive labor movement will be the raising of the unemployment problem to a political issue. The outcome of this will be a further extension of public control.

Political prophecy is a risky affair and I would not venture into details. But large and basic economic currents cannot be ignored for any great length of time, and they are working in one direction. There are many indications that the reactionary tendencies which have controlled American politics are coming to a term. The inequitable distribution of income will bring to the fore the use of taxing power to effect redistribution by means of larger taxation of swollen income and by heavier death duties on large fortunes. The scandal of private appropriation of socially produced values in unused land cannot forever remain unconcealed. The situation in world production and commerce is giving "protection and free trade" totally new meanings. The connection of municipal mismanagement and corruption with special favors to big economic interests, and the connection of the alliance thus formed with crime, are becoming more generally recognized. Local labor bodies are getting more and more discontented with the policy of political abstention and with the farce of working through parties controlled by adverse interests. The movement is cumulative and includes convergence to a common head of many now isolated factors. When a focus is reached, economic issues will be openly and not merely covertly political. The problem of social control of industry and the use of governmental agencies for constructive social ends will become the avowed centre of political struggle.

A chapter is devoted to the political phase of the situation not because it is supposed the place of definitely political action in the resolution of the present split in life is fundamental. But it is accessory. A certain amount of specific change in legislation and administration is required in order to supply the conditions under which other changes may take place in non-political ways. Moreover, the psychological effect of law and political discussion is enormous. Political action provides large-scale models that react into the formation of ideas and ideals about all social matters. One sure way in which the individual who is politically lost, because of the loss of objects to which his loyalties can attach themselves, could recover a composed mind, would be by apprehension of the realities of industry and finance as they function in public and political life. Political apathy such as has marked our thought for many years past is due fundamentally to mental confusion arising from lack of consciousness of any vital connection between politics and daily affairs. The parties have been eager accomplices in maintaining the confusion and unreality. To know where things are going and why they are is to have the material out of which stable objects of purpose and loyalty may be formed. To perceive clearly the actual movement of events is to be on the road to intellectual clarity and order.

The chief value of political reference is that politics so well exemplify the existing social confusion and its causes. The various expressions of public control to which reference has been made have taken place sporadically and in response to the pressure of distressed groups so large that their voting power demanded attention. They have been improvised to meet special occasions. They have not been adopted as parts of any general social policy. Consequently their real import has not been considered; they have been treated as episodic exceptions. We live politically from hand to mouth. Corporate forces are strong enough to secure attention and action now and then, when some emergency forces them upon us, but acknowledgment of them does not inspire consecutive policy. On the other hand, the older individualism is still sufficiently ingrained to obtain allegiance in confused sentiment and in vocal utterance. It persists to such an extent that we can maintain the illusion that it regulates our political thought and behavior. In actuality, appeal to it serves to perpetuate the current disorganization in which financial and industrial power, corporately organized, can deflect economic consequences away from the advantage of the many to serve the privilege of the few.

I know of no recent event so politically interesting as President Hoover's calling of industrial conferences after the stock-market crash of 1929. It is indicative of many things, some of them actual, some of them dimly and ambiguously possible. It testifies to the disturbance created when the prospect of an industrial depression faces a party and administration that have assumed responsibility for prosperity through having claimed credit for it. It testifies to the import of the crowd psychology of suggestion and credulity in American life. Christian Science rules American thought in business affairs; if we can be led to think that certain things do not exist, they perforce have not happened. These conferences also give evidence of our national habit of planlessness in social affairs, of locking the barn-door after the horse has been stolen. For nothing was done until after a crash which every economist—except those hopelessly committed to the doctrine of a "new economic era"—knew was certain to happen, however uncertain they may have been as to its time.

The more ambiguous meaning of these conferences is connected with future developments. It is clear that one of their functions was to add up columns of figures to imposing totals, with a view to their effect on the public imagination. Will there be more than a psychological and arithmetical outcome? A hopeful soul may take it as the beginning of a real application of the engineering mind to social life in its economic phase. He may persuade himself that it is the commencement of the acceptance of social responsibility on a large scale by American industrialists, financiers and politicians. He may envisage a permanent Economic Council finally growing out of the holding of a series of conferences, a council which shall take upon itself a planned coordination of industrial development. He may be optimistic enough to anticipate a time when representatives of labor will meet on equal terms, not for the sake of obtaining a pledge to abstain from efforts to obtain a rise of wages and from strikes, but as an integral factor in maintaining a planned regulation of the bases of national welfare.

The issue is still in the future and uncertain. What is not uncertain is that any such move would, if carried through, mark the acknowledged end of the old social and political epoch and its dominant philosophy. It would be in accord with the spirit of American life if the movement were undertaken by voluntary agreement and endeavor rather than by governmental coercion. There is that much enduring truth in our individualism. But the outcome would surely involve the introduction of social responsibility into our business system to such an extent that the doom of an exclusively pecuniary-profit industry would follow. A coordinating and directive council in which captains of industry and finance would meet with representatives of labor and public officials to plan the regulation of industrial activity would signify that we had entered constructively and voluntarily upon the road which Soviet Russia is traveling with so much attendant destruction and coercion. While, as I have already said, political action is not basic, concentration of attention upon real and vital issues such as attend the public control of industry and finance for the sake of social values would have vast intellectual and emotional reverberations. No phase of our culture would remain unaffected. Politics is a means, not an end. But thought of it as a means will lead to thought of the ends it should serve. It will induce consideration of the ways in which a worthy and rich life for all may be achieved. In so doing, it will restore directive aims and be a significant step forward in the recovery of a unified individuality.

I have tried to make a brief survey of the possibilities of the political situation in general, and not to make either a plea or a prophecy of special political alignments. But any kind of political regeneration within or without the present parties demands first of all a frank intellectual recognition of present tendencies. In a society so rapidly becoming corporate, there is need of associated thought to take account of the realities of the situation and to frame policies in the social interest. Only then can organized action in behalf of the social interest be made a reality. We are in for some kind of socialism, call it by whatever name we please, and no matter what it will be called when it is realized. Economic determinism is now a fact, not a theory. But there is a difference and a choice between a blind, chaotic and unplanned determinism, issuing from business conducted for pecuniary profit, and the determination of a socially planned and ordered development. It is the difference and the choice between a socialism that is public and one that is capitalistic.

7. The Crisis in Culture

Discussion of the state and prospects of American culture abounds. But "culture" is an ambiguous word. With respect to one of its meanings I see no ground for pessimism. Interest in art, science and philosophy is not on the wane; the contrary is the case. There may have been individuals superior in achievement in the past, but I do not know of any time in our history when so many persons were actively concerned, both as producers and as appreciators, with these culminating aspects of civilization. There is a more lively and more widespread interest in ideas, in critical discussion, in all that forms an intellectual life, than ever before. Anyone who can look back over a span of thirty or forty years must be conscious of the difference that a generation has produced. And the movement is going forward, not backward.

About culture in the sense of cultivation of a number of persons, a number on the increase rather than the decrease, I find no ground for any great solicitude. But "culture" has another meaning. It denotes the type of emotion and thought that is characteristic of a people and epoch as a whole, an organic intellectual and moral quality. Without raising the ambiguous question of aristocracy, one can say without fear of denial that a high degree of personal cultivation at the top of society can coexist with a low and unworthy state of culture as a pervasive manifestation of social life. The marvelous achievement of the novel, music and the drama in the Russia of the Tsar's day sufficiently illustrates what is meant. Nor is preoccupation with commerce and wealth an insuperable bar to a flourishing culture. One may cite the fact that the highest phase of Dutch painting came in a time of just

[First published as "The Crisis in Culture. The Fifth Article in Professor Dewey's Series, 'Individualism, Old and New,'" in *New Republic* 62 (19 March 1930): 123–26.]

such expansion. And so it was with the Periclean, Augustan and Elizabethan ages. Excellence of personal cultivation has often, and perhaps usually, been coincident with the political and economic dominance of a few and with periods of material expansion.

I see no reason why we in the United States should not also have golden ages of literature and science. But we are given to looking at this and that "age" marked with great names and great productivity, while forgetting to ask about the roots of the efflorescence. Might it not be argued that the very transitoriness of the glory of these ages proves that its causes were sporadic and accidental? And in any case, a question must be raised as to the growth of native culture in our own country. The idea of democracy is doubtless as ambiguous as is that of aristocracy. But we cannot evade a basic issue. Unless an avowedly democratic people and an undeniably industrial time can achieve something more than an "age" of high personal cultivation, there is something deeply defective in its culture. Such an age would be American in a topographical sense, not in a spiritual one.

This fact gives significance to the question so often raised as to whether the material and mechanistic forces of the machine age are to crush the higher life. In one sense I find, as I have already said, no special danger. Poets, painters, novelists, dramatists, philosophers, scientists, are sure to appear and to find an appreciative audience. But the unique fact about our own civilization is that if it is to achieve and manifest a characteristic culture, it must develop, not on top of an industrial and political substructure, but out of our material civilization itself. It will come by turning a machine age into a significantly new habit of mind and sentiment, or it will not come at all. A cultivation of a class that externally adorns a material civilization will at most merely repeat the sort of thing that has transiently happened many times before.

The question, then, is not merely a quantitative one. It is not a matter of an increased number of persons who will take part in the creation and enjoyment of art and science. It is a qualitative question. Can a material, industrial civilization be converted into a distinctive agency for liberating the minds and refining the emotions of all who take part in it? The cultural question is a political and economic one before it is a definitely cultural one.

It is a commonplace that the problem of the relation of mechanistic and industrial civilization to culture is the deepest and most urgent problem of our day. If interpreters are correct in saying that "Americanization" is becoming universal, it is a problem of the world and not just of our own country—although it is first acutely experienced here. It raises issues of the widest philosophic import. The question of the relation of man and nature, of mind and matter, assumes its vital significance in this context. A "humanism" that separates man from nature will envisage a radically different solution of the industrial and economic perplexities of the age than the humanism entertained by those who find no uncrossable gulf or fixed gap. The former will inevitably look backward for direction; it will strive for a cultivated élite supported on the backs of toiling masses. The latter will have to face the question of whether work itself can become an instrument of culture and of how the masses can share freely in a life enriched in imagination and esthetic enjoyment. This task is set not because of sentimental "humanitarianism," but as the necessary conclusion of the intellectual conviction that while man belongs in nature and mind is connected with matter, humanity and its collective intelligence are the means by which nature is guided to new possibilities.

Many European critics openly judge American life from the standpoint of a dualism of the spiritual and material, and deplore the primacy of the physical as fatal to any culture. They fail to see the depth and range of our problem, which is that of making the material an active instrument in the creation of the life of ideas and art. Many American critics of the present scene are engaged in devising modes of escape. Some flee to Paris or Florence; others take flight in their imagination to India, Athens, the Middle Ages or the American age of Emerson, Thoreau and Melville. Flight is solution by evasion. Return to a dualism consisting of a massive substratum of the material upon which are erected spiritually ornamented façades, is flatly impossible, except upon the penalty of the spiritual disenfranchisement of those permanently condemned to toil mechanically at the machine.

That the cultural problem must be reached through economic roads is testified to by our educational system. No nation has ever been so actively committed to universal schooling as are the people of the United States. But what is our system for? What ends does it serve? That it gives opportunity to many who would otherwise lack it is undeniable. It is also the agency of important welding and fusing processes. These are conditions of creation of a mind that will constitute a distinctive type of culture. But they are conditions only. If our public-school system merely turns out efficient industrial fodder and citizenship fodder in a state controlled by pecuniary industry, as other schools in other nations have turned out efficient cannon fodder, it is not helping to solve the problem of building up a distinctive American culture; it is only aggravating the problem. That which prevents the schools from doing their educational work freely is precisely the pressure—for the most part indirect, to be sure—of domination by the money-motif of our economic regime. The subject is too large to deal with here. But the distinguishing trait of the American student body in our higher schools is a kind of intellectual immaturity. This immaturity is mainly due to their enforced mental seclusion; there is, in their schooling, little free and disinterested concern with the underlying social problems of our civilization. Other typical evidence is found in the training of engineers. Thorstein Veblen—and many others have since repeated his idea—pointed out the strategic position occupied by the engineer in our industrial and technological activity. Engineering schools give excellent technical training. Where is the school that pays systematic attention to the potential social function of the engineering profession?

I refer to the schools in connection with this problem of American culture because they are the formal agencies for producing those mental attitudes, those modes of feeling and thinking, which are the essence of a distinctive culture. But they are not the ultimate formative force. Social institutions, the trend of occupations, the pattern of social arrangements, are the finally controlling influences in shaping minds. The immaturity nurtured in schools is carried over into life. If we Americans manifest, as compared with those of other countries who have had the benefits of higher schooling, a kind of infantilism, it is because our own schooling so largely evades serious consideration of the deeper issues of social life; for it is only through induction into realities that mind can be matured. Consequently the effective education, that which really leaves a stamp on character and

[handwritten marginalia: underlying social problems of our civilization]

thought, is obtained when graduates come to take their part in the activities of an adult society which put exaggerated emphasis upon business and the results of business success. Such an education is at best extremely one-sided; it operates to create the specialized "business mind," and this, in turn, is manifested in leisure as well as in business itself. The one-sidedness is accentuated because of the tragic irrelevancy of prior schooling to the controlling realities of social life. There is little preparation to induce either hardy resistance, discriminating criticism, or the vision and desire to direct economic forces in new channels.

If, then, I select education for special notice, it is because education—in the broad sense of formation of fundamental attitudes of imagination, desire and thinking—is strictly correlative with culture in its inclusive social sense. It is because the educative influence of economic and political institutions is, in the last analysis, even more important than their immediate economic consequences. The mental poverty that comes from one-sided distortion of mind is ultimately more significant than poverty in material goods. To make this assertion is not to gloss over the material harshness that exists. It is rather to point out that under present conditions these material results cannot be separated from development of mind and character. Destitution on the one side and wealth on the other are factors in the determination of that psychological and moral constitution which is the source and the measure of attained culture. I can think of nothing more childishly futile, for example, than the attempt to bring "art" and esthetic enjoyment externally to the multitudes who work in the ugliest surroundings and who leave their ugly factories only to go through depressing streets to eat, sleep and carry on their domestic occupations in grimy, sordid homes. The interest of the younger generation in art and esthetic matters is a hopeful sign of the growth of culture in its narrower sense. But it will readily turn into an escape mechanism unless it develops into an alert interest in the conditions which determine the esthetic environment of the vast multitudes who now live, work and play in surroundings that perforce degrade their tastes and that unconsciously educate them into desire for any kind of enjoyment as long as it is cheap and "exciting."

It is the work of sociologists, psychologists, novelists, dramatists and poets to exhibit the consequences of our present economic regime upon taste, desire, satisfaction and standards of value. An article like this cannot do a work which requires many volumes. But a paragraph suffices to call attention to one central fact. Most of those who are engaged in the outward work of production and distribution of economic commodities have no share—imaginative, intellectual, emotional—in directing the activities in which they physically participate.

It was remarked in an earlier chapter that there is definite restriction placed upon existing corporateness. It is found in the fact that economic associations are fixed in ways which exclude most of the workers in them from taking part in their management. The subordination of the enterprises to pecuniary profit reacts to make the workers "hands" only. Their hearts and brains are not engaged. They execute plans which they do not form, and of whose meaning and intent they are ignorant—beyond the fact that these plans make a profit for others and secure a wage for themselves. To set forth the consequences of this fact upon the experience and the minds of uncounted multitudes would again require volumes. But there is an undeniable limitation of opportunities, and minds are warped, frustrated, unnourished by their activities—the ultimate source of all constant nurture of the spirit. The philosopher's idea of a complete separation of mind and body is realized in thousands of industrial workers, and the result is a depressed body and an empty and distorted mind.

There are instances, here and there, of the intellectual and moral effects which accrue when workers can employ their feelings and imaginations as well as their muscles in what they do. But it is still impossible to foresee in detail what would happen if a system of cooperative control of industry were generally substituted for the present system of exclusion. There would be an enormous liberation of mind, and the mind thus set free would have constant direction and nourishment. Desire for related knowledge, physical and social, would be created and rewarded; initiative and responsibility would be demanded and achieved. One may not, perhaps, be entitled to predict that an efflorescence of a distinctive social culture would immediately result. But one can say without hesitation that we shall attain only the personal cultivation of a class, and not a characteristic American culture, unless this condition is fulfilled. It is impossible for a highly industrialized society to attain a widespread high excellence of mind when multitudes are excluded from occasion for the use of thought and emotion in their daily occupations. The contradiction is so great and so pervasive that a favorable issue is hopeless. We must wrest our general culture from an industrialized civilization; and this fact signifies that industry must itself become a primary educative and cultural force for those engaged in it. The conception that natural science somehow sets a limit to freedom, subjecting men to fixed necessities, is not an intrinsic product of science. Just as with the popular notion that art is a luxury, whose proper abode is the museum and gallery, the notion of literary persons (including some philosophers) that science is an oppression due to the material structure of nature, is ultimately a reflex of the social conditions under which science is applied so as to reach only a pecuniary fruition. Knowledge takes effect in machinery and in the minds of technical directors, but not in the thoughts of those who tend the machines. The alleged fatalism of science is in reality the fatalism of the pecuniary order in which science is employed.

If I have emphasized the effect upon the wage workers, it is not because the consequences are not equally marked with respect to the few who now enjoy the material emoluments of the system and monopolize its management and control. There will doubtless always be leaders, those who will have the more active and leading share in the intellectual direction of great industrial undertakings. But as long as the direction is more concerned with pecuniary profit than with social utility, the resulting intellectual and moral development will be one-sided and warped. An inevitable result of a cooperatively shared control of industry would be the recognition of final use or consumption as the criterion of valuation, decision and direction. When the point of view of consumption is supreme in industry, the latter will be socialized, and I see no way of securing its genuine socialization save as industry is viewed and conducted from the standpoint of the user and enjoyer of services and commodities. For then human values will control economic values. Moreover, as long as means are kept separate from human ends (the consequences produced in human living), "values in use" will be so dominated by exchange or sale values that the former will be interpreted by means of the latter. In other words, there is now no inherent criterion for consumption-values. "Wealth," as Ruskin so vehemently pointed out, includes as much *illth* as well-being. When values in use are the ends of industry they will receive a scrutiny and criticism for which there is no foundation at present, save external moralizing and exhortation. Production for private profit signifies that any kind of consumption will be stimulated that leads to private gain.

There can be no stable and balanced development of mind and character apart from the assumption of responsibility. In an industrialized society, that responsibility must for the most part be associated with industry, since it will grow indirectly out of industry even for those not engaged in it. The wider and fuller the sense of social consequences—that is, of the effect on the life-experience of the consumer—the deeper and surer, the more stable, is the intelligence of those who have the foremost place in the direction of industry. A society saturated with industrialism may evolve a class of highly cultivated persons in the traditional sense of cultivation. But there will be something thin and meagre about even this meed of cultivation if it evolves in isolation from the main currents of action in which thought and desire are engaged. As long as imagination is concerned primarily with obtaining pecuniary success and enjoying its material results, the type of culture will conform to these standards.

Everywhere and at all times the development of mind and its cultural products have been connate with the channels in which mind is exercised and applied. This fact defines the problem of creating a culture that will be characteristically our own. Escape from industrialism on the ground that it is unesthetic and brutal can win only a superficial and restricted success of esteem. It is a silly caricature to interpret such statements as meaning that science should devote itself directly to solving industrial problems, or that poetry and painting should find their material in machines and in machine processes. The question is not one of idealizing present conditions in esthetic treatment, but of discovering and trying to realize the conditions under which vital esthetic production and esthetic appreciation may take place on a generous social scale.

And similarly for science; it is not in the least a matter of considering this and that particular practical application to be squeezed out of science; we have a great deal of that sort of thing already. It is a question of acknowledgment on the part of scientific inquirers of intellectual responsibility; of admitting into their consciousness a perception of what science has actually done, through its counterpart technologies, in making the world and life what they are. This perception would bear fruit by raising the question of what science can do in making a different sort of world and society. Such a science would be at the opposite pole to science conceived as merely a means to special industrial ends. It would, indeed, include in its scope all the technological aspects of the latter, but it would also be concerned with control of their social effects. A humane society would use scientific method, and intelligence with its best equipment, to bring about human consequences. Such a society would meet the demand for a science that is humanistic, and not just physical and technical. "Solutions" of the problem of the relation of the material and the spiritual, of the ideal and the actual, are merely conceptual and at best prophetic unless material conditions are idealized by contributing to cultural consequences. Science is a potential tool of such a liberating spiritualization; the arts, including that of social control, are its fruition.

I do not hold, I think, an exaggerated opinion of the influence that is wielded by so-called "intellectuals"—philosophers, professional and otherwise, critics, writers and professional persons in general having interests beyond their immediate callings. But their present position is not a measure of their possibilities. For they are now intellectually dispersed and divided; this fact is one aspect of what I have called "the lost individual." This internal dissolution is necessarily accompanied by a weak social efficacy. The chaos is due, more than to anything else, to mental withdrawal, to the failure to face the realities of industrialized society. Whether the ultimate influence of the distinctively intellectual or reflective groups is to be great or small, an initial move is theirs. A consciously directed critical consideration of the state of present society in its causes and consequences is a pre-condition of projection of constructive ideas. To be effective, the movement must be organized. But this requirement does not demand the creation of a formal organization; it does demand that a sense of the need and opportunity should possess a sufficiently large number of minds. If it does, the results of their inquiries will converge to a common issue.

This point of view is sometimes represented as a virtual appeal to those primarily engaged in inquiry and reflection to desert their studies, libraries and laboratories and engage in works of social reform. That representation is a caricature. It is not the abandonment of thinking and inquiry that is asked for, but more thinking and more significant inquiry. This "more" is equivalent to a conscious direction of thought and inquiry, and direction can be had only by a realization of problems in the rank of their urgency. The "clerk" and secretary once occupied, if we may trust history, places of great influence if not of honor. In a society of military and political leaders who were illiterate, they must have done much of the thinking and negotiating for which the names of the great now receive credit. The intellectuals of the present are their descendants. Outwardly they have been emancipated and have an independent position formerly lacking. Whether their actual efficacy has been correspondingly increased may be doubted. In some degree, they have attained their liberty in direct ratio to their distance from the scenes of action. A more intimate connection would not signify, I repeat, a surrender of the business of thought, even speculative thought, for the sake of getting busy at some so-called practical matter. Rather would it signify a focussing of thought and intensifying of its quality by bringing it into relation with issues of stupendous meaning.

I am suspicious of all attempts to erect a hierarchy of values: their results generally prove to be inapplicable and abstract. But there is at every time a hierarchy of problems, for there are some issues which underlie and condition others. No one person is going to evolve a constructive solution for the problem of humanizing industrial civilization, of making it and its technology a servant of human life—a problem which is once more equivalent, for us, to that of creating a genuine culture. But general guidance of serious intellectual endeavor by a consciousness of the problem would enable at least one group of individuals to recover a social function and so refind themselves. And recovery by those with special intellectual gifts and equipment from their enforced social defection is at least a first step in a more general reconstruction that will bring integration out of disorder.

Accordingly, I do not wish my remarks about escape and withdrawal to be interpreted as if they were directed at any special group of persons. The flight of particular individuals is symptomatic of the seclusion of existing science, intelligence and art.

The personal gap which, generally speaking, isolates the intellectual worker from the wage earner is symbolic and typical of a deep division of functions. This division is the split between theory and practice in actual operation. The effects of the split are as fatal to culture on the one side as on the other. It signifies that what we call our culture will continue to be, and in increased measure, a survival of inherited European traditions, and that it will not be indigenous. And if it is true, as some hold, that with the extension of machine technology and industrialism the whole world is becoming "Americanized," then the creation of an indigenous culture is no disservice to the traditional European springs of our spiritual life. It will signify, not ingratitude, but the effort to repay a debt.

The solution of the crisis in culture is identical with the recovery of composed, effective and creative individuality. The harmony of individual mind with the realities of a civilization made outwardly corporate by an industry based on technology does not signify that individual minds will be passively molded by existing social conditions as if the latter were fixed and static. When the patterns that form individuality of thought and desire are in line with actuating social forces, that individuality will be released for creative effort. Originality and uniqueness are not opposed to social nurture; they are saved by it from eccentricity and escape. The positive and constructive energy of individuals, as manifested in the remaking and redirection of social forces and conditions, is itself a social necessity. A new culture expressing the possibilities immanent in a machine and material civilization will release whatever is distinctive and potentially creative in individuals, and individuals thus freed will be the constant makers of a continuously new society.

It was said in an earlier chapter that "acceptance" of conditions has two very different meanings. To this statement may now be added the consideration that "conditions" are always moving; they are always in transition to something else. The important question is whether intelligence, whether observation and reflection, intervenes and becomes a directive factor in the transition. The moment it does intervene, conditions become conditions of forecasting consequences; when these consequences present themselves in thought, preference and volition, planning and determination, come into play. To foresee consequences of existing conditions is to surrender neutrality and drift; it is to take sides in behalf of the consequences that are preferred. The cultural consequences that our industrial system now produces have no finality about them. When they are observed and are related discriminatingly to their causes, they become conditions for planning, desiring, choosing. Discriminating inquiry will disclose what part of present results is the outcome of the technological factors at work and what part is due to a legal and economic system which it is within the power of man to modify and transform. It is indeed foolish to assume that an industrial civilization will somehow automatically, from its own inner impetus, produce a new culture. But it is a lazy abdication of responsibility which assumes that a genuine culture can be achieved except first by an active and alert intellectual recognition of the realities of an industrial age, and then by planning to use them in behalf of a significantly human life. To charge that those who urge intellectual acknowledgment or acceptance as the first necessary step stop at this point, and thus end with an optimistic rationalization of the present as if it were final, is a misconstruction that indicates a desire to shirk responsibility for undertaking the task of reconstruction and direction. Or else it waits upon a miracle to beget the culture which is desired by all serious minds.

8.   Individuality in Our Day

In the foregoing chapters, I have attempted to portray the split between the idea of the individual inherited from the past and the realities of a situation that is becoming increasingly corporate. Some of the effects produced on living individuality by this division have been indicated. I have urged that individuality will again become integral and vital when it creates a frame for itself by attention to the scene in which it must perforce exist and develop. It is likely that many persons will regard my statement of the problem as a commonplace. Others will deplore my failure to offer a detailed solution and a definite picture of just what an individual would be if he were in harmony with the realities of American civilization. Still others will think that a disease has been described as a remedy; that the articles are an indiscriminate praise of technological science and of a corporate industrial civilization; that they are an effort to boost upon the bandwagon those reluctant to climb.

I have indeed attempted analysis, rather than either a condemnation of the evils of present society or a recommendation of fixed ends and ideals for their cure. For I think that serious minds are pretty well agreed as to both evils and ideals—as long as both are taken in general terms. Condemnation is too often only a way of displaying superiority; it speaks from outside the scene; it discloses symptoms but not causes. It is impotent to produce; it can only reproduce its own kind. As for ideals, all agree that we want the good life, and that the good life involves freedom and a taste that is trained to appreciate the honorable, the true and the beautiful. But as long as we limit ourselves to generalities, the phrases that express ideals may be transferred

[First published as "Individuality in Our Day. The Sixth and Final Article in Professor Dewey's Series, 'Individualism, Old and New,'" in *New Republic* 62 (2 April 1930): 184–88.]

from conservative to radical or vice versa, and nobody will be the wiser. For, without analysis, they do not descend into the actual scene nor concern themselves with the generative conditions of realization of ideals.

There is danger in the reiteration of eternal verities and ultimate spiritualities. Our sense of the actual is dulled, and we are led to think that in dwelling upon ideal goals we have somehow transcended existing evils. Ideals express possibilities; but they are genuine ideals only in so far as they are possibilities of what is now moving. Imagination can set them free from their encumbrances and project them as a guide in attention to what now exists. But, save as they are related to actualities, they are pictures in a dream.

I have, then, ventured to suppose that analysis of present conditions is of primary importance. Analysis of even a casual kind discloses that these conditions are not fixed. To accept them intellectually is to perceive that they are in flux. Their movement is not destined to a single end. Many outcomes may be projected, and the movement may be directed by many courses to many chosen goals, once conditions have been recognized for what they are. By becoming conscious of their movements and by active participation in their currents, we may guide them to some preferred possibility. In this interaction, individuals attain an integrated being. The individual who intelligently and actively partakes in a perception that is a first step in conscious choice is never so isolated as to be lost nor so quiescent as to be suppressed.

One of the main difficulties in understanding the present and apprehending its human possibilities is the persistence of stereotypes of spiritual life which were formed in old and alien cultures. In static societies—those which the industrial revolution has doomed—acquiescence had a meaning, and so had the projection of fixed ideals. Things were so relatively settled that there was something to acquiesce in, and goals and ideals could be imagined that were as fixed in their way as existing conditions in theirs. The medieval legal system could define "just" prices and wages, for the definition was a formulation of what was customary in the local community; it operated merely to prevent exorbitant deviations. It could prescribe a system of definite duties for all relations, for there was a hierarchical order, and occasions for the exercise of duty fell within an established and hence known order. Communities were local; they did not merge, overlap and interact in all kinds of subtle and hidden ways. A common church was the guardian and administrator of spiritual and ideal truth, and its theoretical authority had direct channels for making itself felt in the practical details of life. Spiritual realities might have their locus in the next world, but this after-world was intimately tied into all the affairs of this world by an institution existing here and now.

To-day there are no patterns sufficiently enduring to provide anything stable in which to acquiesce, and there is no material out of which to frame final and all-inclusive ends. There is, on the other hand, such constant change that acquiescence is but a series of interrupted spasms, and the outcome is mere drifting. In such a situation, fixed and comprehensive goals are but irrelevant dreams, while acquiescence is not a policy but its abnegation.

Again, the machine is condemned wholesale because it is seen through the eyes of a spirituality that belonged to another state of culture. Present evil consequences are treated as if they were eternally necessary, because they cannot be made consistent with the ideals of another age. In reality, a machine age is a challenge to generate new conceptions of the ideal and the spiritual. Ferrero has said that machines "are the barbarians of modern times, which have destroyed the fairest works of ancient civilisations." But even the barbarians were not immutably barbarous; they, too, were bearers of directive movement, and in time they wrought out a civilization that had its own measure of fairness and beauty.

Most attacks on the mechanistic character of science are caused by the survival of philosophies and religions formed when nature was the grim foe of man. The possibility of the present, and therefore its problem, is that through and by science, nature may become the friend and ally of man. I have rarely seen an attack on science as hostile to humanism which did not rest upon a conception of nature formed long before there was any science. That there is much at any time in environing nature which is indifferent and hostile to human values is obvious to any serious mind. When natural knowledge was hardly existent, control of nature was impossible. Without power of control, there was no recourse save to build places of refuge in which man could live in imagination, although not in fact. There is no need to deny the grace and beauty of some of these constructions. But when their imaginary character is once made apparent, it is futile to suppose that men can go on living and sustaining life by them. When they are appealed to for support, the possibilities of the present are not perceived, and its constructive potentialities remain unutilized.

In reading many of the literary appreciations of science, one would gather that until the rise of modern science, men had not been aware that living in nature entails death and renders fortune precarious and uncertain; "science" is even treated as if it were responsible for the revelation of the fact that nature is often a foe of human interests and goods. But the very nature of the creeds that men have entertained in the past and of the rites they have practiced is proof that men were overwhelmingly conscious of this fact. If they had not been, they would not have resorted to magic, miracles, myth and the consolations and compensations of another world and life. As long as these things were sincerely believed in, dualism, anti-naturalism, had a meaning, for the "other world" was then a reality. To surrender the belief and retain the dualism is temporarily possible for bewildered minds. It is a condition which it is impossible to maintain permanently. The alternative is to accept what science tells us of the world in which we live and to resolve to use the agencies it puts within our power to render nature more amenable to human desire and more contributory to human good. "Naturalism" is a word with all kinds of meanings. But a naturalism which perceives that man with his habits, institutions, desires, thoughts, aspirations, ideals and struggles, is within nature, an integral part of it, has the philosophical foundation and the practical inspiration for effort to employ nature as an ally of human ideals and goods such as no dualism can possibly provide.

There are those who welcome science provided it remain "pure"; they see that as a pursuit and contemplated object it is an addition to the enjoyed meaning of life. But they feel that its applications in mechanical inventions are the cause of many of the troubles of modern society. Undoubtedly these applications have brought new modes of unloveliness and suffering. I shall not attempt the impossible task of trying to strike a net balance of ills and enjoyments between the days before and after the practical use of science. The significant point is that application is still restricted. It touches our dealings with things but not with one another. We use scientific method in directing physical but not human energies. Consideration of the full application of science must accordingly be prophetic rather than a record of what has already taken place. Such prophecy is not however without foundation. Even as things are there is a movement in science which foreshadows, if its inherent promise be carried out, a more humane age. For it looks forward to a time when all individuals may share in the discoveries and thoughts of others, to the liberation and enrichment of their own experience.

No scientific inquirer can keep what he finds to himself or turn it to merely private account without losing his scientific standing. Everything discovered belongs to the community of workers. Every new idea and theory has to be submitted to this community for confirmation and test. There is an expanding community of cooperative effort and of truth. It is true enough that these traits are now limited to small groups having a somewhat technical activity. But the existence of such groups reveals a possibility of the present—one of the many possibilities that are a challenge to expansion, and not a ground for retreat and contraction.

Suppose that what now happens in limited circles were extended and generalized. Would the outcome be oppression or emancipation? Inquiry is a challenge, not a passive conformity; application is a means of growth, not of repression. The general adoption of the scientific attitude in human affairs would mean nothing less than a revolutionary change in morals, religion, politics and industry. The fact that we have limited its use so largely to technical matters is not a reproach to science, but to the human beings who use it for private ends and who strive to defeat its social application for fear of destructive effects upon their power and profit. A vision of a day in which the natural sciences and the technologies that flow from them are used as servants of a humane life constitutes the imagination that is relevant to our own time. A humanism that flees from science as an enemy denies the means by which a liberal humanism might become a reality.

The scientific attitude is experimental as well as intrinsically communicative. If it were generally applied, it would liberate us from the heavy burden imposed by dogmas and external standards. Experimental method is something other than the use of blow-pipes, retorts and reagents. It is the foe of every belief that permits habit and wont to dominate invention and discovery,

there isn't a static truth

and ready-made system to override verifiable fact. Constant revision is the work of experimental inquiry. By revision of knowledge and ideas, power to effect transformation is given us. This attitude, once incarnated in the individual mind, would find an operative outlet. If dogmas and institutions tremble when a new idea appears, this shiver is nothing to what would happen if the idea were armed with the means for the continuous discovery of new truth and the criticism of old belief. To "acquiesce" in science is dangerous only for those who would maintain affairs in the existing social order unchanged because of lazy habit or self-interest. For the scientific attitude demands faithfulness to whatever is discovered and steadfastness in adhering to new truth.

The "given" which science calls upon us to accept is not fixed; it is in process. A chemist does not study the elements in order to bow down before them; ability to produce transformations is the outcome. It is said, and truly, that we are now oppressed by the weight of science. But why? Some allowance has to be made, of course, for the time it takes to learn the uses of new means and to appropriate their potentialities. When these means are as radically new as is experimental science, the time required is correspondingly long. But aside from this fact, the multiplication of means and materials is an increase of opportunities and purposes. It marks a release of individuality for affections and deeds more congenial to its own nature. Even the derided bathtub has its individual uses; an individual is not perforce degraded because he has the chance to keep himself clean. The radio will make for standardization and regimentation only as long as individuals refuse to exercise the selective reaction that is theirs. The enemy is not material commodities, but the lack of the will to use them as instruments for achieving preferred possibilities. Imagine a society free from pecuniary domination, and it becomes self-evident that material commodities are invitations to individual taste and choice, and occasions for individual growth. If human beings are not strong and steadfast enough to accept the invitation and take advantage of the proffered occasion, let us put the blame where it belongs.

There is at least this much truth in economic determinism. Industry is not outside of human life, but within it. The genteel tradition shuts its eyes to this fact; emotionally and intellectually it pushes industry and its material phase out into a region remote from human values. To stop with mere emotional rejection and moral condemnation of industry and trade as materialistic is to leave them in this inhuman region where they operate as the instruments of those who employ them for private ends. Exclusion of this sort is an accomplice of the forces that keep things in the saddle. There is a subterranean partnership between those who employ the existing economic order for selfish pecuniary gain and those who turn their backs upon it in the interest of personal complacency, private dignity, and irresponsibility.

Every occupation leaves its impress on individual character and modifies the outlook on life of those who carry it on. No one questions this fact as respects wage-earners tied to the machine, or business men who devote themselves to pecuniary manipulations. Callings may have their roots in innate impulses of human nature but their pursuit does not merely "express" these impulses, leaving them unaltered; their pursuit determines intellectual horizons, precipitates knowledge and ideas, shapes desire and interest. This influence operates in the case of those who set up fine art, science, or religion as ends in themselves, isolated from radiation and expansion into other concerns (such radiation being what "application" signifies) as much as in the case of those who engage in industry. The alternatives are lack of application with consequent narrowing and overspecialization, and application with enlargement and increase of liberality. The narrowing in the case of industry pursued apart from social ends is evident to all thoughtful persons. Intellectual and literary folks who conceive themselves devoted to pursuit of pure truth and uncontaminated beauty too readily overlook the fact that a similar narrowing and hardening takes place in them. Their goods are more refined, but they are also engaged in acquisition; unless they are concerned with use, with expansive interactions, they too become monopolists of capital. And the monopolization of spiritual capital may in the end be more harmful than that of material capital.

The destructive effect of science upon beliefs long cherished and values once prized is, and quite naturally so, a great cause of dread of science and its applications in life. The law of inertia holds of the imagination and its loyalties as truly as of physical things. I do not suppose that it is possible to turn suddenly from these negative effects to possible positive and constructive ones.

But as long as we refuse to make an effort to change the direction in which imagination looks at the world, as long as we remain unwilling to reexamine old standards and values, science will continue to wear its negative aspect. Take science (including its application to the machine) for what it is, and we shall begin to envisage it as a potential creator of new values and ends. We shall have an intimation, on a wide and generous scale, of the release, the increased initiative, independence and inventiveness, which science now brings in its own specialized fields to the individual scientist. It will be seen as a means of originality and individual variation. Even to those sciences which delight in calling themselves "pure," there is a significant lesson in the instinct that leads us to speak of Newton's and Einstein's law.

Because the free working of mind is one of the greatest joys open to man, the scientific attitude, incorporated in individual mind, is something which adds enormously to one's enjoyment of existence. The delights of thinking, of inquiry, are not widely enjoyed at the present time. But the few who experience them would hardly exchange them for other pleasures. Yet they are now as restricted in quality as they are in the number of those who share them. That is to say, as long as "scientific" thinking confines itself to technical fields, it lacks full scope and varied material. Its subject-matter is technical in the degree in which application in human life is shut out. The mind that is hampered by fear lest something old and precious be destroyed is the mind that experiences fear of science. He who has this fear cannot find reward and peace in the discovery of new truths and the projection of new ideals. He does not walk the earth freely, because he is obsessed by the need of protecting some private possession of belief and taste. For the love of private possessions is not confined to material goods.

It is a property of science to find its opportunities in problems, in questions. Since knowing is inquiring, perplexities and difficulties are the meat on which it thrives. The disparities and conflicts that give rise to problems are not something to be dreaded, something to be endured with whatever hardihood one can command; they are things to be grappled with. Each of us experiences these difficulties in the sphere of his personal relations, whether in his more immediate contacts or in the wider associations conventionally called "society." At present, personal frictions are one of the chief causes of suffering. I do not say all suffering would disappear with the incorporation of scientific method into individual disposition; but I do say that it is now immensely increased by our disinclination to treat these frictions as problems to be dealt with intellectually. The distress that comes from being driven in upon ourselves would be largely relieved; it would in part be converted into the enjoyment that attends the free working of mind, if we took them as occasions for the exercise of thought, as problems having an objective direction and outlet.

We all experience, as I have said, the perplexities that arise in the intimacies of personal intercourse. The more remote relations of society also present their troubles. There is much talk of "social problems." But we rarely treat them as problems in the intellectual sense of that word. They are thought of as "evils" needing correction; as naughty or diabolic things to be "reformed." Our preoccupation with these ideas is proof of how far we are from taking the scientific attitude. I do not say that the attitude of the physician who regards his patient as a "beautiful case" is wholly ideal. But it is more wholesome and more promising than the persistence of the pre-scientific habit of anxious concerns with evils and their reform. The current way of treating criminality and criminals is, for example, reminiscent of the way in which diseases were once thought of and dealt with. Their origin was once believed to be moral and personal; some enemy, diabolic or human, was thought to have injected some alien substance or force into the person who was ailing. The possibility of effective treatment began when diseases were regarded as having an intrinsic origin in interactions of the organism and its natural environment. We are only just beginning to think of criminality as an equally intrinsic manifestation of interactions between an individual and the social environment. With respect to it, and with respect to so many other evils, we persist in thinking and acting in prescientific "moral" terms. This pre-scientific conception of "evil" is probably the greatest barrier that exists to that real reform which is identical with constructive remaking.

Because science starts with questions and inquiries it is fatal to all social system-making and programs of fixed ends. In spite of the bankruptcy of past systems of belief, it is hard to surrender our faith in system and in some wholesale belief. We continually reason as if the difficulty were in the particular system that has failed and as if we were on the point of now finally hitting upon one that is true as all the others were false. The real trouble is with the attitude of dependence upon any of them. Scientific method would teach us to break up, to inquire definitely and with particularity, to seek solutions in the terms of concrete problems as they arise. It is not easy to imagine the difference which would follow from the shift of thought to discrimination and analysis. Wholesale creeds and all-inclusive ideals are impotent in the face of actual situations; for doing always means the doing of something in particular. They are worse than impotent. They conduce to blind and vague emotional states in which credulity is at home, and where action, following the lead of overpowering emotion, is easily manipulated by the self-seekers who have kept their heads and wits. Nothing would conduce more, for example, to the elimination of war than the substitution of specific analysis of its causes for the wholesale love of "liberty, humanity, justice and civilization."

All of these considerations would lead to the conclusion that depression of the individual is the individual's own liability, were it not for the time it takes for a new principle to make its way deeply into individual mind on a large scale. But as time goes on, the responsibility becomes an individual one. For individuality is inexpugnable and it is of its nature to assert itself. The first move in recovery of an integrated individual is accordingly with the individual himself. In whatever occupation he finds himself and whatever interest concerns him, he is himself and no other, and he lives in situations that are in some respect flexible and plastic.

We are given to thinking of society in large and vague ways. We should forget "society" and think of law, industry, religion, medicine, politics, art, education, philosophy—and think of them in the plural. For points of contact are not the same for any two persons, and hence the questions which the interests and occupations pose are never twice the same. There is no contact so immutable that it will not yield at some point. All these callings and concerns are the avenues through which the world acts upon us and we upon the world. There is no society at large, no business in general. Harmony with conditions is not a single and monotonous uniformity, but a diversified affair requiring individual attack.

Individuality is inexpugnable because it is a manner of distinctive sensitivity, selection, choice, response and utilization of conditions. For this reason, if for no other, it is impossible to develop integrated individuality by any all-embracing system or program. No individual can make the determination for anyone else; nor can he make it for himself all at once and forever. A native manner of selection gives direction and continuity, but definite expression is found in changing occasions and varied forms. The selective choice and use of conditions have to be continually made and remade. Since we live in a moving world and change with our interactions in it, every act produces a new perspective that demands a new exercise of preference. If, in the long run, an individual remains lost, it is because he has chosen irresponsibility; and if he remains wholly depressed it is because he has chosen the course of easy parasitism.

Acquiescence, in the sense of drifting, is not something to be achieved; it is something to be overcome, something that is "natural" in the sense of being easy. But it assumes a multitude of forms, and Rotarian applause for present conditions is only one of these forms. A different form of submission consists in abandoning the values of a new civilization for those of the past. To assume the uniform of some dead culture is only another means of regimentation. True integration is to be found in relevancy to the present, in active response to conditions as they present themselves, in the effort to make them over according to some consciously chosen possibility.

Individuality is at first spontaneous and unshaped; it is a potentiality, a capacity of development. Even so, it is a unique manner of acting in and with a world of objects and persons. It is not something complete in itself, like a closet in a house or a secret drawer in a desk, filled with treasures that are waiting to be bestowed on the world. Since individuality is a distinctive way of feeling the impacts of the world and of showing a preferential bias in response to these impacts, it develops into shape and form only through interaction with actual conditions; it is no more complete in itself than is a painter's tube of paint without relation to a canvas. The work of art is the truly individual thing; and it is the result of the interaction of paint and canvas through the medium of the artist's distinctive vision and power. In its determination, the potential individuality of the artist takes on visible and enduring form. The imposition of individuality as something made in advance always gives evidence of a mannerism, not of a manner. For the latter is something original and creative; something formed in the very process of creation of other things.

The future is always unpredictable. Ideals, including that of a new and effective individuality, must themselves be framed out of the possibilities of existing conditions, even if these be the conditions that constitute a corporate and industrial age. The ideals take shape and gain a content as they operate in remaking conditions. We may, in order to have continuity of direction, plan a program of action in anticipation of occasions as they emerge. But a program of ends and ideals if kept apart from sensitive and flexible method becomes an encumbrance. For its hard and rigid character assumes a fixed world and a static individual; and neither of these things exists. It implies that we can prophesy the future—an attempt which terminates, as someone has said, in prophesying the past or in its reduplication.

The same Emerson who said that "society is everywhere in conspiracy against its members" also said, and in the same essay, "accept the place the divine providence has found for you, the society of your contemporaries, the connection of events." Now, when events are taken in disconnection and considered apart from the interactions due to the selecting individual, they conspire against individuality. So does society when it is accepted as something already fixed in institutions. But "the connection of events," and "the society of your contemporaries" as formed of moving and multiple associations, are the only means by which the possibilities of individuality can be realized.

Psychiatrists have shown how many disruptions and dissipations of the individual are due to his withdrawal from reality into a merely inner world. There are, however, many subtle forms of retreat, some of which are erected into systems of philosophy and are glorified in current literature. "It is in vain," said Emerson, "that we look for genius to reiterate its miracles in the old arts; it is its instinct to find beauty and holiness in new and necessary facts, in the field and road-side, in the shop and mill." To gain an integrated individuality, each of us needs to cultivate his own garden. But there is no fence about this garden: it is no sharply marked-off enclosure. Our garden is the world, in the angle at which it touches our own manner of being. By accepting the corporate and industrial world in which we live, and by thus fulfilling the pre-condition for interaction with it, we, who are also parts of the moving present, create ourselves as we create an unknown future.

Progressive Education

In the construct of the Democratic society, it becomes everyone's problem

personal troubles as social troubles

draw that linkage between individual psychology and social history

GREAT BOOKS IN PHILOSOPHY PAPERBACK SERIES

ETHICS

Aristotle—*The Nicomachean Ethics*	$8.95
Marcus Aurelius—*Meditations*	5.95
Jeremy Bentham—*The Principles of Morals and Legislation*	8.95
John Dewey—*The Moral Writings of John Dewey, Revised Edition*	
(edited by James Gouinlock)	11.95
Epictetus—*Enchiridion*	4.95
Immanuel Kant—*Fundamental Principles of the Metaphysic of Morals*	5.95
John Stuart Mill—*Utilitarianism*	5.95
George Edward Moore—*Principia Ethica*	8.95
Friedrich Nietzsche—*Beyond Good and Evil*	8.95
Plato—*Protagoras, Philebus,* and *Gorgias*	7.95
Bertrand Russell—*Bertrand Russell On Ethics, Sex, and Marriage*	
(edited by Al Seckel)	19.95
Arthur Schopenhauer—*The Wisdom of Life* and *Counsels and Maxims*	7.95
Benedict de Spinoza—*Ethics* and *The Improvement of the Understanding*	9.95

SOCIAL AND POLITICAL PHILOSOPHY

Aristotle—*The Politics*	7.95
Francis Bacon—*Essays*	6.95
Mikhail Bakunin—*The Basic Bakunin: Writings, 1869–1871*	
(translated and edited by Robert M. Cutler)	11.95
Edmund Burke—*Reflections on the Revolution in France*	7.95
John Dewey—*Freedom and Culture*	10.95
John Dewey—*Individualism Old and New*	9.95
G. W. F. Hegel—*The Philosophy of History*	9.95
G. W. F. Hegel—*Philosophy of Right*	9.95
Thomas Hobbes—*The Leviathan*	7.95
Sidney Hook—*Paradoxes of Freedom*	9.95
Sidney Hook—*Reason, Social Myths, and Democracy*	11.95
John Locke—*Second Treatise on Civil Government*	5.95
Niccolo Machiavelli—*The Prince*	5.95
Karl Marx (with Friedrich Engels)—*The German Ideology,*	
including *Theses on Feuerbach and Introduction to the*	
Critique of Political Economy	10.95
Karl Marx—*The Poverty of Philosophy*	7.95
Karl Marx/Friedrich Engels—*The Economic and Philosophic Manuscripts of 1844*	
and *The Communist Manifesto*	6.95
John Stuart Mill—*Considerations on Representative Government*	6.95
John Stuart Mill—*On Liberty*	5.95
John Stuart Mill—*On Socialism*	7.95
John Stuart Mill—*The Subjection of Women*	5.95
Friedrich Nietzsche—*Thus Spake Zarathustra*	9.95
Thomas Paine—*Common Sense*	6.95
Thomas Paine—*Rights of Man*	7.95
Plato—*Lysis, Phaedrus,* and *Symposium*	6.95
Plato—*The Republic*	9.95
Jean-Jacques Rousseau—*The Social Contract*	5.95
Mary Wollstonecraft—*A Vindication of the Rights of Men*	5.95
Mary Wollstonecraft—*A Vindication of the Rights of Women*	6.95

METAPHYSICS/EPISTEMOLOGY

Aristotle—*De Anima*	6.95
Aristotle—*The Metaphysics*	9.95
George Berkeley—*Three Dialogues Between Hylas and Philonous*	5.95
René Descartes—*Discourse on Method* and *The Meditations*	6.95
John Dewey—*How We Think*	10.95
John Dewey—*The Influence of Darwin on Philosophy and Other Essays*	11.95
Epicurus—*The Essential Epicurus: Letters, Principal Doctrines,*	
Vatican Sayings, and Fragments	
(translated, and with an introduction, by Eugene O'Connor)	5.95
Sidney Hook—*The Quest for Being*	11.95
David Hume—*An Enquiry Concerning Human Understanding*	6.95
David Hume—*Treatise of Human Nature*	9.95
William James—*The Meaning of Truth*	11.95
William James—*Pragmatism*	7.95
Immanuel Kant—*Critique of Practical Reason*	7.95
Immanuel Kant—*Critique of Pure Reason*	9.95
Gottfried Wilhelm Leibniz—*Discourse on Method* and the *Monadology*	6.95
John Locke—*An Essay Concerning Human Understanding*	9.95
Charles S. Peirce—*The Essential Writings*	
(edited by Edward C. Moore, preface by Richard Robin)	10.95
Plato—*The Euthyphro, Apology, Crito,* and *Phaedo*	5.95
Bertrand Russell—*The Problems of Philosophy*	8.95
George Santayana—*The Life of Reason*	9.95
Sextus Empiricus—*Outlines of Pyrrhonism*	8.95

PHILOSOPHY OF RELIGION

Marcus Tullius Cicero—*The Nature of the Gods* and *On Divination*	6.95
W. K. Clifford—*The Ethics of Belief and Other Essays*	
(introduction by Timothy J. Madigan)	6.95
Ludwig Feuerbach—*The Essence of Christianity*	8.95
David Hume—*Dialogues Concerning Natural Religion*	5.95
John Locke—*A Letter Concerning Toleration*	5.95
Lucretius—*On the Nature of Things*	7.95
John Stuart Mill—*Three Essays on Religion*	7.95
Thomas Paine—*The Age of Reason*	13.95
Bertrand Russell—*Bertrand Russell On God and Religion* (edited by Al Seckel)	19.95

ESTHETICS

Aristotle—*The Poetics*	5.95
Aristotle—*Treatise on Rhetoric*	7.95

GREAT MINDS PAPERBACK SERIES

ECONOMICS

Charlotte Perkins Gilman—*Women and Economics: A Study of the*	
Economic Relation between Women and Men	11.95
John Maynard Keynes—*The General Theory of Employment, Interest, and Money*	11.95
Thomas R. Malthus—*An Essay on the Principle of Population*	14.95
Alfred Marshall—*Principles of Economics*	11.95
David Ricardo—*Principles of Political Economy and Taxation*	10.95
Adam Smith—*Wealth of Nations*	9.95
Thorstein Veblen—*Theory of the Leisure Class*	11.95

RELIGION

Thomas Henry Huxley—*Agnosticism and Christianity and Other Essays*	10.95
Ernest Renan—*The Life of Jesus*	11.95
Elizabeth Cady Stanton—*The Woman's Bible*	11.95
Voltaire—*A Treatise on Toleration and Other Essays*	8.95

SCIENCE

Nicolaus Copernicus—*On the Revolutions of Heavenly Spheres*	8.95
Charles Darwin—*The Descent of Man*	18.95
Charles Darwin—*The Origin of Species*	10.95
Albert Einstein—*Relativity*	8.95
Michael Faraday—*The Forces of Matter*	8.95
Galileo Galilei—*Dialogues Concerning Two New Sciences*	9.95
Ernst Haeckel—*The Riddle of the Universe*	11.95
William Harvey—*On the Motion of the Heart and Blood in Animals*	9.95
Werner Heisenberg—*Physics and Philosophy: The Revolution in Modern Science* (introduction by F. S. C. Northrop)	11.95
Julian Huxley—*Evolutionary Humanism*	10.95
Edward Jenner—*Vaccination against Smallpox*	5.95
Johannes Kepler—*Epitome of Copernican Astronomy* and *Harmonies of the World*	8.95
Isaac Newton—*The Principia*	14.95
Louis Pasteur and Joseph Lister—*Germ Theory and Its Application to Medicine* and *On the Antiseptic Principle of the Practice of Surgery*	7.95
Alfred Russel Wallace—*Island Life*	16.95

HISTORY

Edward Gibbon—*On Christianity*	9.95
Herodotus—*The History*	13.95
Thucydides—*History of the Peloponnesian War*	15.95
Andrew D. White—*A History of the Warfare of Science with Theology in Christendom*	19.95

SOCIOLOGY

Emile Durkheim—*Ethics and the Sociology of Morals* (translated with an introduction by Robert T. Hall)	8.95

CRITICAL ESSAYS

Desiderius Erasmus—*The Praise of Folly*	9.95
Jonathan Swift—*A Modest Proposal and Other Satires* (with an introduction by George R. Levine)	8.95
H. G. Wells—*The Conquest of Tme* (with an introduction by Martin Gardner)	8.95

(Prices subject to change without notice.)

Order Form

Prometheus Books
59 John Glenn Drive • Amherst, New York 14228–2197
Telephone: (716) 691–0133

Phone Orders (24 hours):
Toll free (800) 421–0351 • FAX (716) 691–0137
Email: PBooks6205@aol.com

Ship to: _____

Address _____

_____

City _____

County (*N.Y. State Only*) _____

Telephone _____

Prometheus Acct. # _____

❑ Payment enclosed (or)

Charge to ❑ VISA ❑ MasterCard

A/C: ⬜⬜⬜⬜⬜⬜⬜⬜⬜⬜⬜⬜⬜⬜⬜⬜⬜⬜⬜⬜⬜

Exp. Date _____ / _____

Signature _____